"This is a truly valuable book. Dr Paul Tyson's *Theology and Climate Change* delves deeper into the heart of the matter than most treatments of the subject—not only by sketching the *different* rival theologies of nature at play, not only by pointing out how political these theologies are (be they secular or religious), but most importantly by putting the subject in a context often overlooked: that of the implicit ubiquity of theology as *first philosophy*. Dr Tyson highlights the philosophical underpinnings of developments in intellectual history, as now embedded in our very way of life, that are centralmost to the politically hard problems of adequately addressing our ecological crisis. This book should be illuminating and helpful to both Christians grappling with their own theologies of nature in our times and to environmentalists of no Christian commitments who have never really considered how politically important a serious appreciation of the theological roots of the present crisis really is."

Dr Sotiris Mitralexis, University of Winchester, UK

"I found this an engaging and engaged work that multiply illuminates the inseparability of theology and pressing questions of climate change. Paul Tyson offers us a deeply thoughtful and informative reflection on these questions, and at different levels. He does so in relation to old and new theological resources, as well as in relation to current discussions and contemporary controversies. It is written in a voice that will provoke thought in the non-specialist, as well as meeting exacting demands made by the more specialized. It is an impressive essay on the theological dimensions of ecology, in a manner that expands the voice of theology and insightfully includes ecological concerns within the amplitude of a greater mindfulness. Tyson's work offers us a balance of theological finesse and ecological acumen, with a presiding voice not devoid of salutary common sense. His impressively thoughtful voice is both accessible and penetrating. Warmly recommended."

William Desmond, David Cook Chair in Philosophy, Villanova University, USA; Thomas A.F. Kelly, Visiting Chair in Philosophy, Maynooth University, Ireland, Professor of Philosophy Emeritus, Institute of Philosophy, KU Leuven, Belgium

"The human destruction of nature is the ultimate blasphemy and iconoclasm. It is our refusal of the theophanic itself. And yet perverse and heterodox theologies of creation which deny the theophanic are complicit in this unimaginable evil. Too often, they are endorsed by people who laughably imagine that they are 'Conservatives'. All this and more is exposed in Paul Tyson's admirably clear and succinct new book. It can seriously help Christians to help save the Creation."

John Milbank, University of Nottingham, UK

"In this engaging book Paul Tyson presents a penetrating analysis of the generative role of Western science and Christian theology in promoting the idea of progress as dominion over nature. In a time of increased political ferment between Liberals and Conservatives he aims to reach across the ideological divide by remapping their shared roots in Western Christian thought with the help of a remarkable range of reading from revisionist histories of science and religion to Celtic, Evangelical and Orthodox theologies. This all makes for a concise and stimulating read even for those familiar with the terrain."

*Professor Michael Northcott,
Universitas Gadja Madah, Indonesia*

Theology and Climate Change

Theology and Climate Change examines Progressive Dominion Theology (PDT) as a primary cultural driver of anthropogenic climate change. PDT is a distinctive and Western form of Christian theology out of which the modern scientific revolution and technological modernity arises. Basic attitudes to nature, to instrumental power over nature, and to an understanding of humanity's relationship with nature are a function of the deep theological preconditions of Western modernity. Much of what we like about Western modernity is indebted to PDT at the same time that this tacit cultural theology is propelling us towards climate disaster. This text argues that the urgent need to change the fundamental operational assumptions of our way of life is now very hard for us to do, because secular modernity is now largely unaware of its tacit theological commitments.

Modern consumer society, including the global economy that supports this way of life, could not have the operational signatures it currently has without its distinctive theological origin and its ongoing submerged theological assumptions. Some forms of Christian theology are now acutely aware of this dynamic and are determined to change the modern life-world, from first assumptions up, in order to avert climate disaster. At the same time, other forms of Christian theology—aligned with pragmatic fossil fuel interests—advance climate change scepticism and overtly uphold PDT. Theology is, in fact, crucially integral with the politics of climate change, but this is not often understood in anything more than simplistic and polemically expedient ways in environmental and policy contexts. This text aims to dis-embed climate change politics from polarized and unfruitful slinging matches between Conservatives and Progressives of all or no religious commitments.

This fascinating volume is a must-read for those with an interest in environmental policy concerns and in culturally embedded first-order belief commitments.

Paul Tyson is a Senior Research Fellow at the Institute for Advanced Studies in the Humanities at the University of Queensland. His scholarship works across the sociology of knowledge and philosophical theology. He has particular interests in science and religion and in the political implications of culturally embedded beliefs and practices. His recent books include *Kierkegaard's Theological Sociology* (2019) and *Seven Brief Lessons on Magic* (2019).

Routledge Focus on Religion

Narratives of Faith from the Haiti Earthquake
Religion, Natural Hazards and Disaster Response
Roger P. Abbott and Robert S. White

The Bible, Social Media, and Digital Culture
Peter M. Phillips

Religious Studies and the Goal of Interdisciplinarity
Brent Smith

Visual Thought in Russian Religious Philosophy
Pavel Florensky's Theory of the Icon
Clemena Antonova

American Babylon
Christianity and Democracy Before and After Trump
Philip S. Gorski

Avant-garde Art and Radical Material Theology
A Manifesto
Petra Carlsson Redell

Pandemic, Ecology, and Theology
Perspectives on COVID-19
Edited by Alexander J.B. Hampton

Trump and History
Protestant Reactions to 'Make America Great Again'
Matthew Rowley

Theology and Climate Change
Paul Tyson

For more information about this series, please visit: https://www.routledge.com/Routledge-Focus-on-Religion/book-series/RFR

Theology and Climate Change

Paul Tyson

LONDON AND NEW YORK

First published 2021
by Routledge
2 Park Square, Milton Park, Abingdon, Oxon OX14 4RN

and by Routledge
52 Vanderbilt Avenue, New York, NY 10017

Routledge is an imprint of the Taylor & Francis Group, an informa business

© 2021 Paul Tyson

The right of Paul Tyson to be identified as author of this work has been asserted by him in accordance with sections 77 and 78 of the Copyright, Designs and Patents Act 1988.

All rights reserved. No part of this book may be reprinted or reproduced or utilised in any form or by any electronic, mechanical, or other means, now known or hereafter invented, including photocopying and recording, or in any information storage or retrieval system, without permission in writing from the publishers.

Trademark notice: Product or corporate names may be trademarks or registered trademarks, and are used only for identification and explanation without intent to infringe.

British Library Cataloguing-in-Publication Data
A catalogue record for this book is available from the British Library

Library of Congress Cataloging-in-Publication Data
A catalog record has been requested for this book

ISBN: 978-0-367-56536-7 (hbk)
ISBN: 978-0-367-74401-4 (pbk)
ISBN: 978-1-003-09824-9 (ebk)

Typeset in Times New Roman
by MPS Limited, Dehradun

To Professor David Brunckhorst

Outstanding zoologist, landscape ecologist, policy consultant, institutional leader, Christian, and friend.

Contents

Introduction: On Science, Religion, and Politics 1

1 Theology and Climate Change: An Australian Case Study and the Lynn White Thesis 11

PART I
Could Theology Be the Primary Cause of Climate Change? 21

2 On Theology 23

3 The Pre-Conditions of the West's Progressive Dominion Theology 38

4 Progressive Dominion Theology 51

PART II
Towards Christian Theological Solutions to the Global Ecological Crisis 77

5 Contemporary Christian Theologies of Nature and Climate Change: Roman Catholic, Celtic, Orthodox, Indigenous, and Mainline 79

6 Contemporary Christian Theologies of Nature and Climate Change: Evangelicals 95

7 Climate Change and Political Theology	115
Bibliography	128
Index	138

Introduction
On Science, Religion, and Politics

Theology and climate change is a thorny science and religion concern. In this introduction I will outline the manner in which this text approaches science and religion, followed by how I will treat the political dynamics of theology and ecology, followed by a brief outline of the chapters that constitute the argument of the book. This book aims to be broadly educational as regards how theology and climate change interact, with a particular emphasis on the theological backstory of what makes Western modernity tick. This book is also an argument, so the introduction will tell you what sort of solution to what sort of problem this text seeks to put forward.

On Science and Religion

Our assumptions about what science and religion *are* shapes the way we think about relationships between theology and climate change. What is distinctive about this short treatment is the way in which new developments in science and religion thinking are incorporated into it.

This book is one output of a research program called "The After Science and Religion Project" (ASR).[1] This project thinks about knowledge and faith without assuming that either science or religion can be adequately understood as firmly fixed or ultimately definable categories. That is, the project approaches "science" as Western modernity's present approach to the factual knowledge of natural reality. The word "present" carries considerable weight here. This insistence on understanding the historical context in which contemporary scientific knowledge is situated is crucial. Equally, the project approaches "religion" as Western modernity's present approach to understanding those things that contemporary science either excludes from natural reality or re-defines within naturalistic categories, such as meaning, purpose, value, and God. Again, close attention to the historical framework of our present approach to religion is crucial. This project is also very interested in the way our science and religion entail, engage, and exclude each other in our present context. To recognise the interactive and situated nature of science and religion is not an attack on either scientific or theological truth, but it is the result of two important developments in our understanding.

The first development is an awareness that science and religion are not "natural kinds".[2]

Peter Harrison is a historian of modern science and religion of the first rank. His landmark text, *The Territories of Science and Religion*,[3] shows that the way we now think about the proper domain of religion is no older than the seventeenth century. Perhaps even more surprisingly, Harrison shows us that what we now think of as the proper domain of science does not become decisively delineated from philosophy and Christian theology until the late nineteenth century. That is, science and religion as we now understand them—as separated territories that concern themselves with defined and discrete subject matters—are very recent.

Equally, Harrison shows us that our present way of arranging the "territories" of science and religion is not stable, and will almost certainly continue to evolve and re-configure itself. Significantly—as leading historians and sociologists of science clearly demonstrate[4]—the brief heyday of treating science and religion as trans-historical and natural categories, is over. Intellectually, we are now living *after* science and religion.

The second development is the credible argument that modern Western secularism is itself a theological invention.[5]

The Anglican philosophical theologian John Milbank rose to prominence at the end of the twentieth century through his powerful engagements with postmodern critiques of modernity. Milbank makes a compelling argument that both modern conceptions of secular reason and postmodern critiques of modernity presuppose a set of conceptual innovations that were dreamed up by medieval theologians. That is, Milbank shows us that the very grounds that make it possible for us to isolate a distinctly natural territory of immanent knowledge (science) from a discretely supernatural territory of theological belief (religion) are themselves theological. Milbank also critiques the stream of late medieval Western theology that gives rise to modern secularism. Milbank does not find the underlying theology that produced secular modernity adequate as Christian theology, adequate as the grounds of modern secular reason, or adequate as the grounds of a viable postmodern response to the problems of modernity. Milbank then re-inserts his own "radically orthodox" Christian theology back into the very centre of contemporary social theory. Taking the two giants of late classical and high medieval Western theology, Milbank argues that contemporary Augustinian and Thomistic thinking may well show us a way forward for Western thinking. Using this stance, Milbank seeks to address some of the deep difficulties in integrating meaning, knowing, valuing, and power in the modern Western context. In doing this, Milbank took theology firmly out of the seminary and decisively into the secular domains of social theory, sociology, and philosophy.

Milbank's argument has been well-received in Continental and postmodern philosophical circles, and in contemporary social theory circles. But he has produced a considerable ruckus in Christian theological circles and is rather disdained in Anglophone analytic philosophy circles (as, quite often, are Continentally literate thinkers in general). But whether one likes Milbank's work or not, he has produced a broad acceptance by scholars who work in this area that Western theology is most adequately understood as a prior category to modern Western religion, rather than as something that only has a meaningful existence *within* modern religion. Or, to put this another way, modern Western religion itself is produced by a distinctive set of theological developments, as are those categories outside of a modern understanding of religion that we now call secular—including modern science, technology, and politics. In light of Milbank's work, there is a growing awareness that theology is becoming a public discourse again.

The book you are about to read on theology and climate change takes these two developments for granted. Science and religion are not treated as natural kinds, and theology is treated as "bigger" than what we now typically think of as religion.

The academic grounds of this starting point are solid, and yet this is still a new way of thinking. The reason for this is that delineated and essentialist assumptions about the "natures" of science and religion have—over the past 150 years—become embedded in the modern Western way of life. Sociologically, it is not at all easy for us to even imagine how we might actually live *after* science and religion. So even though scholars who specialize in the history of science and the study of religion deeply appreciate that tight definitions of science and religion are no longer possible, and even though social theorists now appreciate that theology is "bigger" than religion, these insights have remarkably little effect on how we still largely *practice* science and religion. This disconnect between scholarship on the one hand, and practice and cultural assumptions on the other hand, is particularly evident in people who are professionally embedded in scientific, educational, and religious institutions. That is, people whose livelihoods are dependent on universities, labs, tech companies, governments, churches, seminaries, etc., find it very difficult to not treat science and religion as fixed and discrete territories. Thinking outside the conceptual and operational boxes of our modern science and religion territories about climate change and theology remains a difficult enterprise. The difficulty is not firstly intellectual; the real difficulty is located in the assumed practical and imaginative horizons of what sociologists call our life-world.

In this text, we shall attempt to grasp both the theoretical and the life-world problems that thinking about theology and climate change poses, working out of a thought-world much indebted to the writings of Harrison and Milbank and working within a sociology of knowledge understanding of how life-worlds and knowledge categories often construct each other. But there are more things than academic thought-worlds and sociology of knowledge dynamics to understand in the arena of theology and climate change. There is also politics.

The Political Dynamic in Theology and Climate Change

As I write in 2020, it is now the case that the Religious Right and the Republican Party in the United States are somewhat united in maintaining a sceptical outlook on climate change science. This political fact has global implications for international environmental agreements. As Nobel Prize-winning economist Daniel Kahneman has pointed out, the most politically intelligent strategy for environmentalists seeking to gain real traction on mitigating global warming would be to win over Evangelical preachers.[6] But to succeed here, environmentalists would have to appreciate that deeply religious people are in significant regards no different to anyone else—this is a hard pill for secular Progressives to swallow. Kahneman is famous for pointing out that in most real-life situations, people from every walk of life do *not* behave in a simplistic, facts-based, rationally self-interested manner. People are genuinely complex. Kahneman points out that the reasons why the US Religious Right is sceptical about climate change has nothing much to do with scientific evidence itself; if one wants to understand their climate scepticism, one needs to look at other areas than science for the causes. In light of that observation, it seems to me that Progressive environmentalists should take a serious interest in Christian theology if they are going to get anywhere with a very religiously framed political opposition.

This short book is particularly interested in seeking a sociologically and theologically adequate understanding of why this political alliance towards climate change inaction has formed between the Religious Right and the Republican Party. To crack open this kind of nut one needs three things:

1. A nuanced understanding of modern, Western, and particularly Evangelical Christian conceptions of nature and humanity.
2. A historical understanding of the culture war that has been raging between Conservative American Christians and Progressive intellectual and cultural reform movements through the twentieth century, particularly as it concerns science and religion.
3. An awareness of the unconscious theological assumptions embedded in *secular* notions of power, enterprise, and freedom in the United States.

In the culture war between Conservative Evangelicals and Progressive environmentalists, both sides feel deeply embattled and tend to see their opponents as somewhere between misguided and belligerently stupid, and somewhere between wrong and evil.[7] The field is now profoundly polarized such that attempts to be in some regards Conservative and in other regards Progressive, is increasingly difficult to do. "Reaching out"—as the Americans say—to the other side in a search for common ground is now very hard to do. The political problem I hope this book might help solve concerns the attempt to find—through a better understanding of theology and sociology—some common meeting points for both religious Conservatives and secular Progressives on environmental issues. In order to pursue this aim, the stance this book holds treats neither Evangelical Conservatives nor Progressive environmentalists as stupid or evil. The line of my argument will not be to say who is right and who is wrong; it is a more uncomfortable line of reasoning than that. My argument aims to show that both religious Conservatives and secular Progressives have a lot more in common than they think, precisely in where they are *both* wrong. Perhaps surprisingly, I will argue that the place where they are both wrong is the *same* theological terrain.

Even so, when it comes to the present state of climate science, I do not think matters of right and wrong can be avoided. For though I fully appreciate the philosophical and sociological complexity of modern science,[8] I see no way of sensibly doubting the very carefully verified scientific evidence that shows us that anthropogenic climate change is in full swing.[9] Climate science scepticism is taken as scientifically unjustifiable in this text, but this does not mean that the theological drivers of Conservative Evangelical Americans in their culture war with Progressives are treated with contempt. To the contrary, if climate change action is to be disentangled from that culture war, the theological drivers of that war need to be understood particularly carefully, and taken seriously.

Given that the global culture of our times is deeply shaped by Western colonial influence over the past 500 years, and empowered by Western science, technology and patterns of natural resource exploitation, if there is a cultural driver of global ecological crisis, it is Western. This is not an anti-Western observation. Given that the religious underlay of today's secularized and post-Christian West is Christian, if there is a religious driver still shaping the underlying cultural attitudes of Western approaches to nature, it is Christian. This is not an anti-Christian observation. And given that around 100 million Christians in the United States are Evangelicals[10] and that it is from this ecclesial tradition that the Religious Right is largely drawn and that this is politically highly significant in shaping the current US policy landscape on some issues (notably climate change), Evangelical theology is particularly entailed in climate change inaction. This is not an anti-Evangelical observation. So the theology that is particularly relevant to a book on theology and climate change is Christian, then Western, then Evangelical. This is the theological landscape that this book will be particularly concerned with.

The argument of the book is that a particular type of Western and Christian theology that I call Progressive Dominion Theology (PDT) profoundly shapes the modern and Western outlook on nature. PDT, then, is the most basic cultural cause of anthropogenic climate change. This is not a new idea, but it is usually read by Conservatives as an attack on Christian theology, and it is usually read by Progressives as not applicable to them because many of them have explicitly renounced Christian theology. This book argues for a re-understanding of what theology itself is, and for a less polarized reflex to either affirm or denounce the value and meaning of the Christian heritage of Western attitudes to nature. Such an approach to theology and climate change could—it is hoped—open up a constructive common ground as regards the proper care of nature for both Conservatives and Progressives.

An Outline of the Argument

This book has two sections. The first part defends the claim that Progressive Dominion Theology is intimately historically embedded in the deep cultural drivers of climate change, and that PDT remains powerfully active in secularised Western modernity today. Part one will seek to demonstrate that whether people are Christian or not, or even religious or not, theology remains a primary cultural force shaping the practices and values that define our life-world. The second part of the book is an argument proposing a theological solution to climate change via a careful analysis of basically two trajectories in Western Christian theology. The fact that this will be a Christian theological argument is important. For—whether people celebrate or mourn it—the global politico-economic world order that is producing anthropogenic climate change is the progeny of Western European Christian theology. Advocating a better Christian theology of nature than PDT is the obvious place to start trying to address this problem.

The first chapter opens up the problem of theology and climate change by reference to a concrete Australian climate disaster as played out in the devastating national bush fires of 2019–2020. This chapter quickly looks at how theology and politics—even in very secular Australia—play out. Lynn White's Thesis will also be introduced in this chapter, and a brief defence of arguing along White's lines for a "big" conception of theology will be put forward.

Moving further into a "big" idea of theology, Chapter 2 defines theology in Aristotelian terms as first philosophy. Theology here concerns one's most basic primary assumptions, the grounds on which one's understanding of the nature and meaning of reality and one's own humanity, are built. From here we will look at a few basic theological stances on the nature of nature, and the nature of human nature. There aren't that many primary stances to choose from, but differences at this level make an enormous difference in how one understands what nature is and what a healthy relationship with nature should be.

The third chapter looks at three theological innovations in the fourteenth and seventeenth centuries that had to happen before what we now know as modern science could appear. The fourteenth-century Franciscan innovations of nominalism (where only particular individual things exist) and voluntarism (where will, power, and freedom are the most fundamental characteristics of God, in whose image we are created) will be explained. Then we will examine the seventeenth-century idea of what I will call "pure matter" (where physical reality becomes reductively material). "Pure matter" is produced by the recovery of ancient atomism and the rejection of the Aristotelian matter-and-form conceptions of natural beings. This new way of thinking about matter will also be described as a theological innovation with

8 *Theology and Climate Change*

profound implications for the modern age. These explicitly late medieval and early modern theological developments underpin the new natural philosophy—what we now call science—that emerged in the seventeenth century.

The fourth chapter looks more closely at the Christian theological underpinnings that gave rise to modern science and technology. This chapter will look at two developments that emerged in the seventeenth century: 1) a distinctive Christian conception of dominion over nature; and 2) a distinctive Christian concept of progress, as measured in terms of utility, the increase of instrumental power, and the advancement of human flourishing through science and technology. The argument will be put forward that a distinctly Modern, Western, and Christian Progressive Dominion Theology has produced both the modern scientific age (with its many virtues) and anthropogenic climate change (with its catastrophic dangers) in the present.

Secular Western modernity has become increasingly pragmatic and materialistic since the late nineteenth century, and so its public theological intelligence has faded dramatically from view, leaving theological reflexes in place without any substantive interest in what those reflexes used to refer to. But outside of the secularized public context, and within the new religiously defined domain of theology, many Christian theologians have been thinking hard about the theology of nature. An awareness that the technological powers of humanity were becoming damaging to the earth has been growing since the Industrial Revolution, and a considerable re-working of the modern Western theology of nature has been going on.

Chapter 5 briefly looks at five Christian approaches to the theology of nature that have either solidly distanced themselves from, or never been affiliated with, Progressive Dominion Theology. We quickly look at Roman Catholic, Celtic, Orthodox, Indigenous, and Mainline Christian theologies of nature. Together, these show a set of surprisingly unified alternatives to PDT. Because PDT has a specifically Christian genealogy, it is not surprising that Christian theologians have been crunching this terrain over very carefully, particularly over the past 50 years. There is a dynamic for change at play in the underlying theology of nature in the Christian world, which could have radical climate change response implications for our Western-originated global civilization. But theology itself would have to be understood differently, in a less discretely secularised way, if these alternative trajectories to PDT were to have influence. A Progressive determination to keep theology out of the public square can be neatly aligned with a publically pragmatic and privately religious understanding of religious freedom, which works against these non-PDT Christian trajectories gaining political traction in the secular West.

Chapter 6 looks at American Evangelical theology. This type of Christian theology is often strongly invested in various forms of PDT, and yet the picture there is complex. Understanding the intricate relations between PDT and the unconscious theological assumptions of pragmatic, secular, commercial Western power is very important in understanding the political dynamics that keeps PDT rolling on. But there are real potentials for change in that arena, and change in the Evangelical arena may really be the most critical issue for the ecological future of the planet. Conversely, cultural warfare based on embattled Christian Conservatives fighting embattled secular Progressives is the most likely recipe for climate action failure.

Chapter 7 looks at climate change and political theology. It explores two types of political theology: religious political theology (focusing on Evangelicals) and philosophical political theology (exploring the first-order meaning assumptions of secular Western modernity). The conclusion seeks to draw all the threads together in the book to show how an emerging Christian theology of nature may well be able to replace PDT and save the planet.

But now, to something very tangible and immediate: bushfires.

Notes

1 This project is run through the Institute for the Advanced Study of the Humanities at the University of Queensland, Australia. I am very grateful to the Templeton World Charity Foundation, as well as the Issachar Fund, for their broad interest in funding science and religion research and in providing grant money to the University of Queensland to facilitate this project.
2 Harrison, *The Territories of Science and Religion*, 4. "The label 'natural kind' is applied to natural groupings of things, the identity of which is natural in the sense that it does not depend on human beings ... My argument with regard to the categories 'religion' and 'science' is that to some degree we are mistaken in thinking that they are analogous to natural kinds ..."
3 This text is based on Professor Harrison's 2010–2011 Gifford Lectures at the University of Edinburgh.
4 See, for example: Gaukroger, *The Natural and the Human*; Shapin, *Never Pure*; Latour, *Science in Action*.
5 See Milbank, *Theology and Social Theory*. Milbank's text famously opens with the sentence "Once, there was no 'secular'" (p. 9) and goes on to show how secular reason has its origins in theological developments in Western theology that are intimately entangled in fourteenth-century Franciscan innovations. See also – from an anthropological perspective – parallel arguments unpacking

the novelty and distinctiveness of modern Western ways of being both religious and secular: Asad, *Formations of the Secular*; Asad, *Genealogies of Religion*.
6 Kahneman on *Hidden Brain* hosted by Shankar Vedantam, "Daniel Kahneman on Misery, Memory, and Our Understanding of the Mind" National Public Radio, USA, 12 March 2018. Listen to the audio from 24 minutes to 28 minutes. See also, Kahneman, Slovic, and Tversky (eds.) *Judgment under Uncertainty*.
7 Smith, *American Evangelicalism*. George, *Hijacking America*.
8 See Chalmers, *What Is This Thing Called Science?*; Feyerabend, *The Tyranny of Science*; Shapin and Schaffer, *Leviathan and the Air-Pump*; Polanyi, *Personal Knowledge*.
9 For accessible and cogent explanations of the authoritative scientific consensus on anthropogenic climate change, see: Houghton, *Global Warming*; Maslin, *Climate Change*. For detailed and extensive research showing how climate scientists have measured and modelled the globe's climate temperature responses to increased carbon dioxide in the atmosphere see, for example, Sherwood, Webb, Annan, et al., "An Assessment of Earth's Climate Sensitivity Using Multiple Lines of Evidence."
10 Veldman, *The Gospel of Climate Change Skepticism*, 4.

1 Theology and Climate Change
An Australian Case Study and the Lynn White Thesis

1.1 Australia's Worst Bushfire Season, 2019–2020

In January 2020 there were abnormally intense bushfires raging all over the Southern Continent. Many properties and 33 human lives were lost. 12.6 million hectares of bush were destroyed, 431 million tonnes of carbon dioxide were released into the atmosphere, Australia's largest cities were blanketed in smoke, and a Conservative estimate of one billion bush animals died.[1] Professional, volunteer, and military firefighters were hard-pressed, at the same time, all over the country.

Australia—where I live—is prone to fire, but records have never seen a bushfire summer like this.[2] Since British settlement, we have not seen this semi-arid continent dried out and heated up like this, over a period of years, resulting in this crisis.[3] The penny is starting to drop—particularly among the young—that climate change is not some distant fantasy, but a very real and present danger.[4] Australian fire chiefs are also deeply aware of this problem.[5]

Yet, Australia's Prime Minister during the 2019-2020 fire season, Mr Scott Morrison, firmly denied that these fires signalled any climate change policy failure by his or previous Australian governments.[6] This defensive response was not surprising as the 2019 national election was won by Mr Morrison, assisted by the Murdoch press, staunchly backing the opening of what is likely to become the globe's largest open-cut coal mine.[7]

The Morrison government resolutely supports the mining and exporting of Australia's enormous coal deposits, even though there is a clear causal link between carbon dioxide emissions from coal-fired power stations and global warming.[8] A strong alliance between the primary resource sector, the mass media, and Australia's two central political parties has a long back-story in Australia, well predating the twenty-first century.[9] There was nothing particularly unusual when, before the 2019 election, the media was full of stories about environmental extremism and the need for a strong mining sector to provide jobs for Australians. The facts that there will not be many jobs and, as far as the human causes of climate change go, coal-fired power stations are the worst offenders, were irrelevant to the winning political narrative of the 2019 election.[10] On a platform of sound commercial sense, jobs for regional Australia, and national economic interest, the Morrison government never wavered from its firm commitment to fuelling the heavily air-polluting coal-fired power stations of Asia.[11]

Currently (in 2020), over 500,000,000 metric tonnes of coal is mined in Australia each year, all of which is burnt, which clearly *increases* the large amounts of carbon dioxide entering the globe's dangerously overheating atmosphere.[12] Australia produces enough coal to burn each year for more than four Australias,[13] but somehow Australia's substantial export contribution to promoting global climate change does not show up in the Morrison government's unambitious and trickily accounted emissions reduction targets.[14] The huge volume of atmospheric carbon dioxide released by our enormous bushfires does not show up in Australia's emission calculations either.

Regionally, Oceania leaders at the 2019 Pacific Island Forum implored Mr Morrison to ease off the climate change accelerator as low-lying island nations are disappearing into the Pacific already. Mr Morrison, a passionate Pentecostal Christian, told the Islander leaders that whilst he stands by them "as family", he is committed to the economic welfare of Australians before anything else—surely they must understand that.[15]

Mr Morrison shows every sign of a genuine, compassionate concern for traumatised Australians who have lost all their possessions in the 2019–2020 fire season. Yet, he shows no signs of ecological grief when our reefs die, when extinctions multiply, when the arctic melts, and when California, the Amazon basin, and Australia burn.[16] Supporting prevailing economic and commercial necessities are more important to the Australian Prime Minister than any serious attempt to address climate change by actively re-configuring our power generation technologies, let alone reducing, then stopping, Australian coal exports.

The three top messaging and policy priorities for the Morrison government are 1) a strong economy; 2) safety from terrorists, illegal immigrants, and the COVID-19 pandemic; and 3) national pride. Mr Morrison's background in marketing stands him in good stead as a successful political leader. He does not deviate from this messaging in promoting his political brand. As a result, he is committed to providing certainty for the mining sector in Australia as their prosperity (and royalties for government coffers) is the central pillar of his first political priority. Climate change is a public relations concern that he must manage, but it is off the main game to him. Indeed, if it interferes with mining royalties, it is a public relations enterprise that must be firmly positioned beneath economic necessity.

In Australia, religion, as an indicator of Conservative political tendencies, only seems accidentally aligned with the commercial interests of the fossil fuel sector. Prime Minister Morrison happily expresses religious sentiments when communicating his personal convictions, but his sharp political pragmatism, his calculative electoral realism, and his government's close ties to the mining sector, are neatly cordoned off from his personal faith. To Mr Morrison, politics is public and religion is private, so how could theology have any real influence on climate change policy in Australia?

1.2 Can Theology Cause Climate Change?

In 1967, Professor Lynn White Jr. published a short but highly influential paper titled "The Historical Roots of Our Ecological Crisis".[17] White argued that there is a distinctive Western way of thinking about our relation to nature which is causing our present ecological crisis, and that this way is intimately bound up with Christian theology. Clearly, both politicians with deep personal religious convictions and board members of entirely secular natural resource companies would likely find this an incomprehensible claim. However, White cannot be dismissed.

White thought that our technological powers had become so great that we were now placing depletion and despoiling pressures on the natural environment that it may no longer be able to absorb. White did not think we would solve this problem without re-thinking and re-feeling our most fundamental relationship to nature—our theological relationship. We somehow need to reconfigure the operational and imaginative horizons that shape our understanding of who we are, what nature is, what technological power should be used for, and what our relationship to nature should be. To White, as a historian of technology,[18] such a reconfiguring could not be achieved without a close and deep exploration of Western theology.

It is important to recognise that White is not simply *blaming* Christianity, or theology, for our ecological woes—for White does not think culturally embedded theology is optional. Indeed, White's understanding of theology in general, and Christian theology in particular, is both filial and hopeful. White finds it natural to assume that drawing on the rich resources of Christian theology provides us with the best set of ready-at-hand tools to use if we wish to fix the causes of our ecological crisis at its theological roots.[19]

White's short, crisp paper has been immensely influential. It has largely produced the contemporary scholarly field of eco-theology. This field is strikingly inter-disciplinary, combining environmental science, socio-cultural studies, theology, feminism, history, philosophy, biblical studies, and religious studies. Eco-theology, though indebted to White, has steadily grown in sophistication and scholarly range since 1967.[20] By now, an enormous literature has grown up around White's Thesis, so I must make it clear what type of treatment of White's Thesis I will employ in this book. Importantly, I am not approaching White's Thesis as an eco-theologian, but as a sociologist of knowledge.[21]

"Knowledge"—sociologically—is an expansive category. We are all born into a socially situated understanding of reality that tells us the kind of things that are factually true. We also simply absorb those things that our society takes to be normatively true and operationally realistic, and these, too—sociologically—are integral with the knowledge which constitutes our social reality. We don't choose the assumed reality framework we are born into in exactly the same way that we don't choose what language we learn as a child. Having always simply assumed the reality and language we are born into, the first principles that shape the deep values, beliefs, and meanings that underpin our collective way of life are usually invisible to us. The sociologist of knowledge tries to do two things: firstly, to make the underlying assumptions defining the knowledge that is native to any given life-world visible; and secondly, to understand how those knowledge categories shape the parameters of realistic action within that life-world.

Reading White's Thesis as a sociologist of knowledge, three things seem uncontroversially true:

1. Contemporary Western modernity knows nature in a distinctive way, which shapes what is normative to us regarding how we use nature.
2. The deep meaning, belief, and value foundations of Western modernity's present knowledge of nature are tied up with how we understand science and technology, and that understanding is shaped by a deeper backstory of assumed meanings and truths embedded in Western Christian theology.

3. If we cannot re-think our presently assumed first-order knowledge categories, there seems little hope that we will change the entrenched pattern of ecological degradation that is native to our prevailing life-world.

Taking these three insights from White as valid, this book will not delve into the expansive literature in eco-theology concerning what other aspects of White's Thesis are right or wrong. Instead, reading White in a sociology of knowledge manner, this book offers a brief defence of the following startling two-pronged claim:

A. A particular type of Christian theology is the primary underlying cause of contemporary anthropogenic climate change; and
B. Christian theology still provides us with the best starting place to seek to respond to this crisis.

The above is a startling claim because White thinks that whether you personally are religious or not, a certain type of Christian theology is the most primary definer of your culturally assumed approach to the meaning and use of nature. White has what I will call a "big" and substrata-located view of what theology is, which ignores secular/religious separations, and which does not situate theology exclusively within a religious domain. Theology—in White's "big" usage—is something much more pervasive and deep-seated than what we typically think of as religion.

Back in the 1960s, White's "big" conception of theology was not (and is still not) an easy or palatable idea for many of White's readers to grasp. Anti-Christian Progressives and Christian Conservatives can be equally offended by, and reject, White's Thesis. Anti-Christian Progressives often believe that science and common-sense materialism rises in triumphant opposition to the Christian religion. For example, the evangelistic atheists of recent memory all seem to believe that they have entirely discarded Christian theology and embraced scientific materialism instead.[22] If White is claiming (which he is) that *all* Western modernists are tacitly embedded in a Christian theological understanding of nature, then our Progressive scientific atheists would beg to differ. Many Conservative Christians, on the other hand, may well admit that Christians have failed to live up to their own creation care theology, but they feel that White is blaming them for the world's ecological problems. Conservative Christians locked in cultural warfare with Progressive Atheists are often inclined to maintain that *if* we have an ecological crisis, it is produced by the casting *off* of a genuinely

Christian theology of nature, rather than by Christian theology's malign influence.[23]

We will look more closely at why the Conservative suspicion of White's Thesis is misplaced in Chapter 6, but we will here quickly touch on why the Progressive suspicion is also misplaced.

Bernard Lightman, a renowned historian of the Victorian era, points out that in the latter half of the nineteenth century, a small group of influential figures in England and America conducted a brilliantly effective campaign to professionalise science.[24] They did this in order to break the deep influence of the church on the universities and to wrest natural philosophy out of the hands of amateurs, many of whom were either women or clergy. The success of this enterprise had far-reaching impacts on our educational institutions and, from there, on Western culture at large. A significant tool developed to promote a new era of strictly secular and professionalised scientific naturalism was the Conflict Thesis. This thesis was a reconstructive "historical" argument that purported to show that religion was the implacable enemy of science, and that progress meant the advance of science and the retreat of religion. The Conflict Thesis has no historical credibility—we would not have Western science without Western religion[25]—yet this "triumph of science over religion" narrative has enjoyed astonishing success in the academy throughout the twentieth century and up to the present. Although it is false, the Conflict Thesis has become the origins myth of our bold new secular age in the academy,[26] and has become an assumed cultural truth underpinning the functional materialism and hedonistic consumerism embraced by the post-1960s cultural revolution. But the historical truth is, White knows what he is talking about: Western modernity—including its science, including twentieth-century atheism, and including post-1960s consumer culture—really is profoundly embedded in and indebted to a particular trajectory of Christian theology.

I do not think that White can be properly read as either pro or anti Progressive, or pro or anti Conservative. His argument remains, I think, equally confronting to all sides of the main antagonists in our contemporary culture wars about science and religion. If White is right, then those culture wars are unwinnable for either side, as the war itself is premised on a historically false conception of the opposition of science and religion. To White, *both* modern science and modern religion are grounded in the *same* basic theological framework. Because I think White has grasped something profound here, that could take a great deal of heat out of these culture wars, I am requesting the reader at this point attempt to think of White's argument in reasonably open and neutral terms, as a hypothesis to be carefully considered.

In line with White's argument, I wish to call the "big" theology underpinning the modern Western approach to nature, *Progressive Dominion Theology* (PDT). Understanding what PDT is, how it shapes our civilizational norms and realistic practices in relation to nature, and how we might reform this theology so that it no longer tilts us towards ecological destruction is the task the rest of this book seeks to perform.

Let us proceed, then, by asking a question that is much more demanding to answer than is normally assumed: what is theology?

Notes

1 Werner, "Wildfires with Wild Numbers: Fact Checking a Catastrophe." *Science Friction*, Australian Broadcasting Corporation, 16 February 2020. Khadem, "Ross Garnaut's Climate Change Prediction Is Coming True and It's Going to Cost Australia Billions, Experts Warn." *Australian Broadcasting Corporation News*, 8 January 2020. See also Garnett, Wintle, Lindenmayer, et al., "Conservation Scientists Are Grieving after the Bushfires – But We Must Not Give Up." *The Conversation*, 21 January 2020. Elsworth, "NSW Bushfires Lead to Deaths of Over a Billion Animals and 'Hundreds of Billions' of Insects, Experts Say." *Australian Broadcasting Corporation News*, 9 January 2020.
2 Griffiths (Australian National University), "Savage Summer." *Inside Story*, 8 January 2020.
3 Nationwide climate change induced heating and drying, resulting in unprecedented bushfire devastation in 2020, was accurately modelled by the Garnaut Climate Change Review back in September 2008. See again Khadem in note 1 above.
4 In September 2019, a nationwide climate strike attracted 300,000 participants, mainly young. See here: "Global Climate Strike Sees 'Hundreds of Thousands' of Australians Rally across the Country." *Australian Broadcasting Corporation News*, 21 September 2019, https://www.abc.net.au/news/2019-09-20/school-strike-for-climate-draws-thousands-to-australian-rallies/11531612.
5 In 23 April 2019, former fire and emergency service leaders set up the "Emergency Leaders for Climate Action Group" to try and get the ear of the Australian government about the climate change-induced catastrophic pending fire season. They produced this statement: https://emergencyleadersforclimateaction.org.au/wp-content/uploads/2019/04/CC_MVSA0184-Firefighting-and-Emergency-Services-Statement-A4-Version_V4-FA.pdf. These retired fire specialists felt it necessary to make this statement because they knew that employed fire chiefs were effectively muzzled by the policy commitments of the Australian government that refused to treat climate change as a real and present threat. For background on the formation of the Emergency Leaders for Climate Change Group, see here: https://www.climatecouncil.org.au/emergency-leaders-climate-action/. Even when the 2019–2020 bushfire season had started, these former leaders were unheeded by the Morrison government.

See here: "Former Fire Chiefs 'Tried to Warn Scott Morrison' to Bring in More Water-Bombers Ahead of Horror Bushfire Season." *Australian Broadcasting Corporation News*, 15 November 2019, https://www.abc.net.au/news/2019-11-14/former-fire-chief-calls-out-pm-over-refusal-of-meeting/11705330.

6 Albeck-Ripka, Tarabay, and Kwai, "As Fires Rage, Australia Sees Its Leader as Missing in Action." *New York Times*, 4 January 2020.

7 A commitment by the Morrison government to green-light the environmentally controversial and enormous open-cut Carmichael coal mine in Queensland was a key election issue in 2019. See Horn, "Election 2019: Why Queensland Turned Its Back on Labor and Helped Scott Morrison to Victory." Australian Broadcasting Corporation, *News*, 24 May 2019. The conservative and rural vote backed the conservative government's promise of coal mining jobs if a very strange deal with the Adani mining company's proposed Carmichael mine was allowed to go ahead. Concerning the bizarre economics of this deal with Adani, see Quiggin, "Explaining Adani." *The Conversation*, 3 June 2019.

8 Foster and Bedrosyan, "Understanding CO_2 Emissions from the Global Energy Sector" The World Bank, briefing paper 85126, 2014/5, p1. http://documents1.worldbank.org/curated/en/873091468155720710/pdf/851260BRI0Live00Box382147B00PUBLIC0.pdf "Coal is, by far, the largest source of energy-related CO_2 emissions globally, accounting for more than 70 percent of the total. This reflects both the widespread use of coal to generate electrical power, as well as the exceptionally high CO_2 intensity of coal-fired power."

9 Hamilton, *Scorcher*.

10 Regarding rapidly increasing global atmospheric CO_2, see Lindsey, "Climate Change: Atmospheric Carbon Dioxide." National Oceanic and Atmospheric Administration, U.S. Department of Commerce. Climate.gov, 19 September 2019. Regarding the primary role coal-fired power stations play in global atmospheric CO_2 emissions, see Davis and Socolow, "Commitment Accounting of CO_2 Emissions," 8 (26 August 2014). Regarding Adani's Carmichael mine and mining jobs, see Bradley, "How Australia's Coal Madness Led to Adani." *The Monthly*, April 2019.

11 Australia's coal exports are currently estimated to be 380,000,000 metric tonnes per annum (September 2019). See Cunningham, Van Uffelen, and Chambers, "The Changing Global Market for Australian Coal." *Reserve Bank of Australia Bulletin*, September 2019, 1. The Reserve Bank notices that the future of Australian exported coal is unclear as global North countries are transitioning to renewables. Even so, global South countries, Asia in particular, are, at present, strongly increasing their consumption of imported coal for coal-fired power stations. And all the time global atmospheric carbon dioxide concentrations continue to rise at an unprecedented speed. See Lindsey in note 10 above. Even so, Australia's coal export market is starting to slip down slightly from a staggering high in 2016. In 2016 Australia was the biggest national net global exporter of coal – 389,000,000 metric tonnes – as well as producing another 114,000,000 metric tonnes for domestic coal-fired power stations and steel production. See Birol, "Key World Energy Statistics 2017." International Energy Agency, 17.

12 Coal production figures as cited in Cunningham et al. in note 11 above. See also Lindsey in note 10 above regarding the unprecedented increase of global atmospheric carbon dioxide from 340 parts per million in 1980 to 407 parts per million in 2018.
13 See note 11 above.
14 On Australia's low international ranking in atmospheric CO_2 emissions reduction, see https://climateactiontracker.org/countries/australia/. On the Morrison government's use of "carryover credits" to appear to meet its very low reduction target for 2020 in 2030, see Merzian, "Taking Way Too Much Credit." The Australia Institute Briefing Note, May 2019. This briefing opens thus: "Australia's emissions have increased every year since 2014, when the Australian Government became the first country to repeal a national carbon pricing system. Government projections from December 2018 show under the current suite of policies national emissions will continue to increase, rather than decrease. The projections also show Australia is not on track to meet its emission reduction target of 26% by 2030 from a 2005 baseline."
15 Cain, "Scott Morrison's Challenge at Pacific Islands Forum in Tuvalu Is to Deliver on Climate Change." *The Conversation, Australian Broadcasting Corporation News*, 16 August 2019.
16 On ecological grief, see Vince and Gatehouse, "Ecological Grief." BBC World Service, 20 January 2020; Gordon, Radford, and Simpson, "Grieving Environmental Scientists Need Support," 193, doi:10.1126/science.aaz2422. On coral bleaching, see https://www.coralcoe.org.au/for-managers/coral-bleaching-and-the-great-barrier-reef. On increasing extinction rates see: 30,000 currently highly endangered species on the International Conservation of Nature Union red list: https://www.iucnredlist.org/. On the melting Arctic Circle see Serreze, *Brave New Arctic*. On California burning and further issues see Klein, *On Fire*. On Australia burning, 126,000 square kilometres (a bit over 31,000,000 acres) were alight in the fire 2019/20 season. This is an area slightly larger than the US state of Mississippi; see Werner, "Wildfires with Wild Numbers: Fact Checking a Catastrophe." Australian Broadcasting Corporation, *Science Friction*, 16 February 2020.
17 White, "The Historical Roots of Our Ecological Crisis."
18 White, *Medieval Technology and Social Change*.
19 White argued that St Francis of Assisi shows the West a form of environmental theology that could save us from our ecological crisis.
20 See, for example, LeVasseur, *Religion and Ecological Crisis*.
21 The classic accessible account of the sociology of knowledge is Berger and Luckmann, *The Social Construction of Reality*.
22 See Dawkins, *The God Delusion*; Hitchens, *God Is Not Great*; Dennett, *Darwin's Dangerous Idea*; Grayling, *The God Argument*; Krauss, *The Greatest Story Ever Told … So Far*.
23 Schaeffer, for example, accepts that there is a theologically caused ecological crisis, but finds White's Thesis being used to justify a pantheistic theology of nature to replace a poor Christian theology of nature. Schaeffer rejects that solution and argues that a good Christian theology of nature – that is, a genuinely biblical and Reformed theology – is the solution to the problem. See Schaeffer, *Pollution and the Death of Man*, 11, 23–25, 39–44.

24 Lightman (ed.), *Rethinking History, Science, and Religion*; Lightman and Reidy (eds.), *The Age of Scientific Naturalism*; Lightman, *Victorian Popularizers of Science*.
25 Harrison, *The Territories of Science and Religion*, 171–175; Numbers, *Galileo Goes to Jail and Other Myths about Science and Religion*.
26 Aechtner, "Galileo Still Goes to Jail."

Part I
Could Theology Be the Primary Cause of Climate Change?

2 On Theology

If you are not religious, then the word "theology" may strike you as referring to pre-modern ideas about supernatural things that have no bearing on your life. Even if you are religious, you may have little interest in theology and think it has nothing much to do with day-to-day life; theology is for the seminarian. This sort of disinterest in theology is entirely understandable because theology has come to have a discretely religious meaning in recent times, and religion itself has come to mean private convictions that are more or less forbidden entry to the practical world of secular public life. Yet, this particular way of thinking about what "religion" itself *is* (and hence, what theology is) is very recent; it is no older than the modern era, and it only really became widely accepted within Western modernity since the late Victorian era.[1] Indeed, typical contemporary attitudes towards religion, theology, and the secular hide more than they reveal. This makes it genuinely hard for us to grasp how significant theology actually is to climate change.

To re-frame the way we think about theology, let us firstly take a quick look at something that is usually thought of as being entirely non-religious: time.

2.1 The West's Theological Understanding of Time

Lynn White wrote another important paper back in 1942, titled "Christian Myth and Christian History".[2] Among other things, he pointed out that the way we understand the nature of time is profoundly influenced by the West's long embedding in Christian belief.

White explains that classical Greco-Roman views of historical time were largely either cyclical or degenerative, or had no direction as such but expressed a purposeless temporal undulation. In contrast, the Christian view of history is linear, singular, and purposive.

White explains:

> The axiom of the uniqueness of the Incarnation[3] required a belief that history is a straight-line sequence guided by God. And as the Church became the exclusive cult of the Roman Empire, the doctrines of undulation and recurrent cycles vanished from the Mediterranean world. No more radical revolution has ever taken place in the world-outlook of a large area ... During the Middle Ages and Renaissance, step by step, this ... providential interpretation [of linear history] was very gradually secularized into the modern idea of progress ...[4]

That is, there is a purposive, progressive, and activist approach to time in Western culture, drawn from its religious formation that simply isn't there in cultures that are shaped by more cyclical, degenerative, undulating, or magical views of time. Western people—whether they happen to have Christian faith or not—are profoundly shaped by the Christian religion in their shared cultural assumptions about the nature of historical time. This gives Western people a conception of cosmically significant purposive action within time, aiming at some as yet unrealised redemptive perfection. We are always going somewhere, always getting better. Atheist Progressives of the late nineteenth century onwards are just as profoundly shaped by the Christian conception of historical time as were early modern thinkers like Francis Bacon to whom advances in science were an obviously Christian means of hurrying towards the fulfilment of history.[5]

In the West, we live our tightly scheduled, progressive, goal-defined lives without a moment's thought to this being a function of our culture's distinctive religious formation. Many non-Western people are not naturally like Westerners in this regard and have learnt how to become frenetic time managers as a result of being drawn into today's global world. For our modern global civilisation itself was created by this linear, forward-driving Western tendency, as manifest in the modern colonial age of European global conquest.

The contemporary Western attitude to time remains linear and purposive, even as the religious tutelage of a sacred cosmic history defined by creation, fall, redemption, and eschaton drops away. Which is to say that Christian theology is more deeply formative of secular Western modernity, and more fundamental to the most basic operational assumptions of our distinctive cultural life-world than we like to acknowledge. And this is White's point, and this is why he later claims (1967) that we will not get a handle on why we have an ecological crisis until we look closely at the underpinning theological bearings that still direct Western culture.

Theology—as understood in this book—is the endeavour to articulate what the deep assumptions shaping the common practices of our culture are, concerning that which is most foundational in our collective value and purpose belief structures. Or, as Aristotle put it, theology is first philosophy.

We shall now have a quick look at Aristotle's understanding of theology, and this will involve a brief argument in favour of continuing to take Aristotle's theological reasoning seriously today.[6] Don't be "religiously" worried about this: Aristotle himself was entirely non-religious by modern standards, and the term "religion" is anachronistic as applied to Aristotle. Even so, Aristotle's understanding that there are certain given truths underpinning any reasoned philosophy of nature, and certain given truths underpinning any sort of understanding about the value, purpose, and meaning of the reality we actually experience, retains a powerful intellectual force. Exploring these givens to try and understand what it is that we simply assume when we think of nature and reality, is what theology—to Aristotle—is all about.

2.2 Theology as First Philosophy

This section is likely to seem quite philosophically laboured, and the connections between the West's Christian theology and climate change are going to be nowhere in sight, at first. This seemingly off-topic slog, however, is unavoidable. Aristotle gives us a way into thinking about theology in a similar register to White, but this is now a register that is structurally hard for us to grasp. Aristotle—by his very distance from modern Western thinking—can help us see what White is talking about.

To Aristotle, theology is first philosophy.[7] Aristotle thinks that a rational and valid understanding of nature and the good life depends on our grasp—however incomplete—of the primary realities of Being, Intelligence, and Goodness.[8] This, Aristotle calls theology, because these primary realities are, to him, divine. The divine, then, is that which is *the grounds of* human knowledge and reason, and concerns that which is *the cause of* order, value, purpose, and meaning in the cosmos.

One must not think of theology, in Aristotle's sense, as being what we now think of as "religious". For Aristotle was not at all religious, as we understand religion. No doubt he would have participated in holidays and cultic ceremonies that were inextricable from the normal civic life of any ancient Greek city-state.[9] But he saw such practices as imaginatively constructed out of the customary tissue of the ancient regulative traditions of the people, and as integral with the familial, political, and commercial structures of Greek life.[10] What we modern people mean by the word "religion" had nothing much to do with divinity to Aristotle. But when it does come to divinity, Aristotle's theology was of a high rationalist tenor.[11] Theologically, he had no interest in cults, miracles, myths, and doctrines about supernatural beings. Indeed, he had considerable disdain for what he considered to be the fantastic tales told by Homer and Hesiod regarding the various obscenities and horrors committed by the gods.

The key thing I want to draw attention to in Aristotle's understanding of theology is his approach to foundational concerns. For in this regard, all theological outlooks are strikingly different to typically modern philosophical outlooks. This is important to grasp as it defines what is distinctive about theology itself, as contrasted with the usual working assumptions of modern philosophy.

To Aristotle, there are certain things that one must take as the *grounds* of reasonable proof, which cannot themselves be proved.[12] He takes it as reasonable to simply accept, in good faith, that our sense perceptions are truth revealing of an ordered, purposive, and intelligible cosmos that our minds can understand. This, he does not think, can be proved *by* sense and logic, but sense and logic are impossible as truth revealing if this is not true. To perceive and reason whilst maintaining good faith in sensory perception and reasoned argument, then, requires a commitment to those foundational truths that make truth revealing perception and valid argument possible. This may seem obvious, but it is precisely this attitude that modern philosophy has sought to reject. Modern philosophy looks for the foundation of true knowledge *within* sense and reason. If we cannot empirically and rationally prove the validity of sense and reason, then we think our knowledge of the world is bereft of any true meaning. To modern philosophy we must have proof—defined *within* the warrants of sense and logic—or there is no real truth in our knowledge of the world. This is the rejection of theology in Aristotle's sense. Let me unpack this a bit.

Classically modern philosophy—particularly from the eighteenth century onwards—increasingly rejects the idea of *believing* first principles and requires instead a *demonstrated proof* of first principles. But—problematically—such first-order proofs have not been forthcoming. Yet, classical philosophical modernists remain deeply committed to rational and empirical proofs and equally determined to reject belief (religious faith and metaphysical speculation) as truth carrying. In the nineteenth century, this resulted in re-defining truth itself in ways that required neither first demonstration, nor metaphysical speculation, nor religious faith. We came to define truth in terms of practical power, probable repeatability, functional coherence, and poetic meaning constructions. Modern "realism" is now (ironically) *grounded* in non-rational, non-purposive, non-meaningful first principles. Our first principles are now the instinctive givens of Nature and the social, political, and economic constructions of Culture. "Realism" today—when one takes it as given that there is no meaning or purpose to reality itself—is about as incompatible with an Aristotelian conception of theology as it is possible to get.

"Realism" today tends to hold that meaning, value, and purpose are cultural glosses that we project onto reality, but reality *itself* has no qualities, purpose, or meaning; reality only has observable quantities and measurable relations of distance, motion, energy, and force. That is, we don't tend to think of Aristotle's qualitative, purposive, and intellectual grounds of human sense and reason as being features of objective reality. Instead, because we tend to think that only things within the purview of science and mathematics can be proven, anything beyond or prior to the world of sense and mathematical descriptions, and anything qualitative and intellective in itself, is just not true. This means we can believe whatever we like (this is our freedom of conscience) about non-demonstrable first principles, but this is because *all* beliefs of such nature are just as speculative as each other, and none of them amounts to knowledge.

In sum, we tend to assume that we do not have—in Aristotle's sense—any foundational theological commitments about why the world itself is meaningful, or concerning how perception and logic are tied in to any divine order. Or, we might think we have *personal* theological beliefs if we are "religious", and we may even be deeply committed to these beliefs. But—unlike Aristotle—we know our theological beliefs are firmly isolated from public knowledge claims, and so they are private convictions that our liberal society allows us to have, as expressions of our personal religious freedoms. Both of these ways of thinking make it very hard for us to grasp what Aristotle means by theology.

But modern realism—as irrealists and postmodernists happily point out—is self-defeating.[13] The claim that reality has no meaning is itself a meaning claim. The claim that intelligibility is not real has, in itself, to be intelligible if we are to take it seriously. And the problem with trying to get rational and empirical truth foundations for all modern knowledge claims—apart from the embarrassing fact that we just can't lift reason and sensation up to the level of first truths by their own boot straps[14]—is that we are linguistic and interpretive beings. In other words, value and meaning do not disappear just because we can't make pure science and pure reason carry them.[15] And we still have to *believe* that value and meaning are just made up—we can't even prove that!

The humbling truth is, Aristotle is still very persuasive about the inescapability of first philosophy (theology). There really are first premises of meaningful thought and valuable action that we simply have to accept in good faith. This is, inescapably, just how the human condition works. For any meaningful system of knowledge, one has to hold certain first truths as reasonable and reliable even though these first truths cannot themselves be demonstrated. To Aristotle, this means we are *given* perceivable and logical meanings as the starting place for understanding the purposeful and intelligible cosmos in which we live. The giver of such first truths is divinity to Aristotle; hence, any system of belief concerning the grounds of meaning, knowledge, and qualitative truth, is theology.

Whether you share Aristotle's confidence that the grounds of human meaning, knowing, and valuing are somehow divine or not, from a sociological perspective, deeply embedded common meanings and values are inescapable.[16] This is quite offensive to our idea of personal belief freedoms concerning our own moral and religious (or irreligious) commitments, but the disturbing truth is that those ideas of personal freedom are also given to us by our modern liberal cultural life-world. And whether God or Culture (or God through Culture) gives us our shared meanings, values, and purposes, we do not really choose these things for ourselves. They are given. Theology is concerned with such givens, and with how they are expressed in the common way of life we share with others who have much the same basic foundational beliefs and outlooks.

Interestingly, most primary values and meanings embedded in any particular cultural life-world are connected to what we would now call religious beliefs. Lynn White's examination of time shows this well. And when it comes to our understanding of nature, this is—just as a matter of history—inescapably religiously formed in one way or another. Hence, despite the embedded prejudices of contemporary secular modernity against the very idea of common religious beliefs being profound shapers of the life-world in which we now live, we will not get anywhere in seriously appreciating our own deep attitudes towards nature without overcoming the "Enlightened" revulsion of religion. Religion remains an unavoidable social and first philosophy reality in the secular, post-Christian West. Let us, then, move on from Aristotle and dive into the religiously formed theological attitudes to nature that underpin the modern secular Western world-view.

2.3 Four Theologies of Nature

To understand the distinctive features of modern Western theologies of nature, some appreciation of the range of entirely different approaches to the philosophy of nature is important. The four basic alternative theologies of nature that will be drawn on in this text are Epicurean, Animist, Christian, and—for want of a better term—Eastern. Epicurean theology is reductively materialist, implies a randomly undulating conception of historical time, and denies any divinely given sense of cosmic meaning. Animist theology correlates with a magical sense of time, there is no meaningful distinction between the material and the spiritual, and cosmic significance is embedded in immanent experience at the same time as time, matter, and space are magical and transcendental. Christian theology correlates with a linear and purposive sense of time, it has—in recent centuries—set up a sharp distinction between an imminent, practical, and material realm, and a supernatural realm. Here, transcendence ultimately defines reality, whereas the immanent is passing away. Eastern theology correlates with a cyclical sense of time, and either finds 'heaven' to be a higher order of the natural, or finds immanent experience to be, ultimately, an illusion. Each theology has quite distinctive understandings of the self, of anthropology, of power and—of course—of nature.

I should make it clear that I do not intend these four broad natural theology families to be all-inclusive (I have left out, for example, Stoic and Taoist theologies of nature), and I certainly do not intend these labels to be in any sense adequate to understanding non-Western theologies of nature. But these labels are important in understanding contemporary thought *within* Western thinking. As it is a distinctive Western theology of nature that this text is most concerned to understand, this limited way of thinking about what the different first philosophy approaches to nature are, is useful for the limited purpose of this text. Let us look, then, in a little more detail at what these four labels signify.

2.3.1 Epicurean Theologies of Nature

The Epicurean theology of nature was a minority and intellectually elite outlook in Greco-Roman antiquity. Four key components of this theology are Epicurus' distinctive hedonistic ethics, the atomism of Democritus, the scepticism of Pyrrho, and (as later developed) the naturalism of Lucretius.

To Epicurus, pleasure and pain are the only determinates of what one should do. That is, there is no transcendent right or wrong that stands above the desires, fears, and necessities of embodied life, so nature alone tells us what is right. A rational ordering of our natural desires, fears, and necessities, avoids unnecessary pain. The acceptance of natural givens should produce a moderated life of realistic expectations as concerns our bodily needs and enjoyments. This hedonism advocates restraint from excess—which always has unwanted health and wellbeing implications—and aims at what we might call the simple pleasures of body, mind, and community, such as the moderated satisfaction of bodily appetites, education, and friendship.[17] Epicurean hedonism, then, has little in common with 1960s style pop culture hedonism, which is largely defined by the casting off of the restraint of culturally Christian morality, and the ardent pursuit of pleasure, come what may as regards to unpleasant and destructive consequences.

To Democritus, the world appears to be populated by living and thinking beings that have knowable and purposive natures, but in fact, this is an illusion. Our conventional knowledge is just an appearance, in reality, there are only three final truths: atoms, motion, and void. Democritus seems to have grasped the truth of an inherently arbitrary atomic reality by some special contemplative insight.[18]

To Pyrrho, certain knowledge based on sensory experience, or any attempt to know the truth via reason, is an illusion.[19]

To Lucretius, the nature of things needs no higher explanation than that which is tangibly presented to the sensory manifold of our experience. Taking the world simply as it is given to us to know it (presuming atomist materialism, Epicurean hedonism, and accepting the sceptical limitations of human knowledge); we need not impose our own views of purpose or meaning on it but can delight in it simply "as it is". Here, nature and existence—let alone our own nature and existence—has no purpose or meaning that comes to it from some transcendent and spiritually meaningful "beyond". We find peace within ourselves when we simply accept nature and ourselves as we are.[20]

Here, nature has no purpose, history has no direction, all human values and meaning are constructions of natural inclinations and imagination, the appearance of order and meaning is simply an appearance, and there is nothing to fear from death as it is simply the end of one's existence (there is no immaterial soul). This is a first philosophy that allows for complete freedom in the construction of meaning and purpose in human concerns, and allows nature to be totally disinterested in human affairs. Here, we have no responsibilities of stewardship towards and no right to rulership over nature. We are entirely a part of nature, which itself is indifferent to us and to itself.

It must be noted here that there is an intimate relationship between this Epicurean theology of nature and contemporary naturalism, for this ancient outlook has a vital life in our times and is the particular theological darling of modern scientistic atheism. This is complex.

Epicurean thinking died out in late antiquity, but was revived, along with an atomist understanding of matter, in waves in the Renaissance, in early modern thinking, and in the nineteenth century. It is integral with the rise of nineteenth-century reductively materialist outlooks associated with Progressive nineteenth-century atheism, though modern scientific naturalism remains a complex hybrid of both Epicurean and Christian first philosophy premises.

A contemporary Epicurean view of nature is typically aligned with a distinctively modern outlook on knowledge called "objectivity".[21] "Objectivity" tries to isolate an analysis of the factual from meanings, values, beliefs, and religious and philosophical commitments, in order to simply observe and understand what is apparent to our immediate sensory perception and to reasonable (mathematical) analysis. It aims at philosophical indifference, at agnosticism with regards to value, purpose, and meaning, and at theoretical impartiality. Of course, to its advocates, "objectivity" is itself a value (objectivity is good) and objective knowledge is thought of as—in at least some sense—truth revealing, so "objectivity" needs to be framed within a larger outlook on value and meaning, and that larger outlook is the theological framework of modern secular naturalism.

It is clearly the case that ancient Epicureanism had a strong influence on the rise of modern science. Even so, Christian theology was a stronger influence probably right up until the 1960s. Trajectories leading to the modern de-magicing of nature, the focus on practical use, and the isolation of the perceivable and only sceptically knowable realm of immediate nature from revealed truth, all have strongly Christian theological roots.[22] So there is a complex blend of potentially (and, by the late nineteenth century, actually) conflicted first philosophy commitments regarding outlooks on nature in Western science. But at this point, all that needs to be grasped is that the original Epicurean Theology of nature in its Classical form has no relationship to the Christian Theology of nature to be outlined shortly.

2.3.2 Animist Theologies of Nature

This outlook is what Western Christendom used to call pagan. Here, nature itself is divine and we, as part of nature, are also magical beings, as are our artefacts. Meaning, purpose, and will are inherent in nature to this outlook, and the mythic/magical world—including inanimate rocks, landscapes, heavenly bodies, and natural forces—is alive and purposive, and all things are embedded in their own songlines in the living fabric of reality.

In recent decades, modern science has made an interesting alliance with aspects of this prescientific outlook. James Lovelock's "Gaia Hypothesis" was an attempt to blend modern science with seeing the planet as a living biosphere. The science of the self-regulating biosphere is fascinating. Lovelock's work has gained considerable popular attention (and perhaps misappropriation) with those modern Western people seeking to recover a way of seeing the world as sacred, and remains important in the environmental movement today.[23] But Lovelock's complex attempt at a scientific animist environmental theology has little real contact with Indigenous non-Western epistemologies that treat the physical and the spiritual in at least a functionally integrated manner.

Time is typically cyclical or undulating in this outlook, as history and creation have no redemptive direction or culminating purpose. Temporal transience and the cycle of birth and death express a recycling vision of nature, which is not a nature where individual souls or even enduring principles or timeless meanings are eternal. This can be very similar to what I will call an Eastern theology of nature, where creation and destruction, light and darkness, good and evil, are both entirely natural and entirely divine, and are one.

2.3.3 Christian Theologies of Nature

Here, creation is not its own origin and does not determine its own destiny. Creation is the work of an all-powerful, all-Good, all-knowing, transcendent God. Divine Reason (*Logos*) is the creative and unifying source of the cosmos. This *Logos* is Good (hence, there is a moral structure to created reality), and divine order, essence, and meaning is gifted to creation from beyond itself. Qualities, intelligible meanings, inherent essences, and innate purposes are real features of creation, not simply human glosses. There is an original harmony in nature that is good. But traditional Christian theologies of nature maintain that this original harmony was disrupted so that the balances and norms of nature under the present natural order are in some deep sense wounded and in need of healing. Thus, creation has a redemptive trajectory that is finally guided by God, but in which humanity is a central player. Human actions, thus, affect nature, and human actors are in some manner placed over nature by God.

I have called this a traditional Christian Theology of Nature, but the "Christian" part of this philosophy of nature is embedded in a broader outlook defined by a transcendent creator God and the concept of an ordered and purposeful cosmos. This "Christian" theology of nature has taken over aspects of prior Abrahamic and Greek *Logos* thinking. But the focus will be on the specifics of the Christian version of this philosophy of nature because that is the version most intimately connected to Western modernity, which is the outlook particularly entangled in the present environmental perils of the planet.

2.3.4 Eastern Theologies of Nature

I must confess to being outside of my intellectual comfort zone here, but there is another philosophy of nature which I will call Eastern that is, perhaps, illustrated to some extent by some forms of Buddhist teaching.[24] Here, the individual soul is an illusion, and the world of space and time is an illusion, and release from the suffering wheel of birth and death into a Nirvana beyond personhood, beyond existence, beyond the flux and contingency of joy and sorrow, is salvation. Nature does not, finally, exist. Attachment to things material and transitory is an attachment to illusions and the cause of all sorrow. Detachment, particularly from the bondages and illusions of ego, is the path of compassion and freedom. That is, there is a form of theology/philosophy that is—shall we say—fully spiritual, which overcomes the trials of material temporal existence by the enlightened realisation that space and time are not ultimately real.

2.4 Nature and Theology

We have now sketched four broad families of first philosophy outlooks on the nature of nature. These we can call basic types of Theologies of Nature. There are several implications that follow from identifying these first philosophies. The first thing to note here is that all cultures have tacit theologies of nature as a simple matter of fact. Secondly, the culture of Western modernity now underpins the common operational assumptions of global commerce, technology, law, and natural resource exploitation which is producing anthropogenic climate change. Thirdly, there is a particular form of a Christian theology of Nature that is embedded in the deep cultural assumptions of Western modernity, though an Epicurean theology of nature is now also prominent in Western modernity. Fourthly, even though these theologies operate at a tacit collective level, and even though much of Western modernity is now post-Christian, a Christian theology of nature remains the most deep-seated civilisational theology of nature embedded within Western modernity. The deep cultural drivers shaping our present common values and meanings as regards our normal approach to the use of nature are theological. Unless we can understand this theology of nature and change at least aspects of it, there is very little prospect of any serious alteration of the course to environmental destruction we are currently on.

Up until the mid-nineteenth century, by far the most widespread theology of nature within Western modernity was Christian. From Francis Bacon to Ludwig Feuerbach, the mainstream of natural philosophy of Western modernity—what we now call modern science—was deeply embedded in an assumed Christian theology of nature. The event horizon that significantly changed this was the powerful intellectual insertion of a scientific neo-Epicurean theology of nature, with its naturalistic materialism, into Western culture. This event horizon was being passed through, intellectually, over most of the nineteenth century, starting in earnest with *The Life of Jesus Critically Examined* by David Strauss[25] in 1836, coming towards the top of its arc with *The Origins of Species* by Charles Darwin in 1859, and somewhat completing that arc with Andrew Dickson White's *History of the Warfare of Science with Theology in Christendom* in 1897. This event horizon is being passed through culturally from roughly the 1860s to the 1960s, largely through the outworking of this intellectual revolution in university education. Looking back to the 1920s, we can see that once the Christian theology of nature no longer had a naturally assumed grasp on the emerging scientific establishment, or on the intellectual culture at large, and once the social and educational reforming agendas of a new Epicurean theology of nature had achieved real impetus in the realm of Western knowledge early in the twentieth century, the door, so to speak, was open for Westernised versions of Animist and Eastern theologies of nature to propagate in the West as well.²⁶ So whilst the deep cultural inertia of more than a thousand years of Christian formation on Western European culture still largely shapes the habits and practices of Western modernity, the actual theological landscape of different approaches to nature is now wide open to change. This recent theological openness makes Conservative Christians seeking to hold onto an explicitly Christian theology of nature, and seeking to uphold Western cultural traditions embedded in that theology, particularly insecure.

There is no understanding of the culture war in which Conservative Christians and Progressive secularists are locked without appreciating the manner in which competing theologies of nature are the most fundamental drivers of this conflict. It is theologies of nature that separate an Evangelical understanding of creation care from what they fear is a neo-pagan, pantheistic, anti-Christian, yet also reductively materialist environmental ideology. Even so—at least according to Lynn White (whom I agree with about this)—the Christian theology of nature remains the most inertial background culture forming frame for contemporary Western culture, even after the arrival of serious rival first-philosophies of nature from the late nineteenth century. To see how this is so, we must (again, like Lynn White) dig back deeper into the West's cultural history; we must go back earlier than the scientific revolution to the medieval grounds out of which the scientific revolution arose. We are going to have to look quite closely at the theology of the Middle Ages if we want to understand how Western Modernity's outlook on nature works, today, at a first philosophy level.

Notes

1 See Harrison, *The Territories of Science and Religion*, 83–116. Since the publication of Wilfred Cantwell Smith's *The Meaning and End of Religion* [1962], the field of Religious Studies has firmly dropped the idea that you can come up with a catch-all trans-historical definition of what religion is. There are plenty of good recent books on this, such as Nongbri, *Before Religion. A History of a Modern Concept*.
2 White, "Christian Myth and Christian History," 145–158.
3 The "Incarnation" is the Christian doctrine that God took up human flesh, within human history, once and only once, in the life of Jesus of Nazareth.
4 White, "Christian Myth and Christian History", 147.
5 See chapter 9 "Restoring Lost Dominion: Bacon and the Millennium" in Henry, *Knowledge Is Power*, 108–119.
6 For a very interesting argument in taking Aristotle seriously today, see Roochnik, *Retrieving Aristotle*.
7 Aristotle, *Metaphysics*, VI, I. 8–12. (1026a, 10–32).

8 See Menn, "Aristotle and Plato on God as Nous and as the Good" for an excellent account of Aristotle's theology of God drawn largely from some close readings from Aristotle's *Physics, Metaphysics*, and *Eudemian Ethics*.
9 The political life of the ancient Greek city state was essentially tied to cultic festivals and public temples. See Price, *Religion of the Ancient Greeks*, 25–46.
10 Aristotle, *Metaphysics*, 1074 b 1–14; and Aristotle, *Politics*, 1314b 39–1315b 4.
11 God, to Aristotle, is "thought thinking itself" (Aristotle, *Metaphysics*, 1074b, 33), the unmoved "prime mover" (Aristotle, *Metaphysics*, 1072b, 10.), the rational ordering source (in a sense accidentally) of the cosmos. Such Deity is the cause of all motion, because the lower is always attracted in love to the higher; but such a high rational divinity has no particular interest in us (being lower). For a somewhat Aristotelian contemporary natural theology, see Davies, *The Mind of God*.
12 This presupposition of primary meanings was mainstream in the ancient ways of thinking about knowledge, certainly within the Platonist and Aristotelian trajectories. See Gerson, *Ancient Epistemology*.
13 See Goodman, *Ways of Worldmaking*; Baudrillard, *Simulacra and Simulation*.
14 The difficulty of relating *a priori* logical reason to *a posteriori* sensory experience in a unified view of demonstrable truth has been a perennial problem for modern philosophy. The circularity of modern rationalism and Hume's problematization of induction have put to an end to the idea that reason and perception could provide their own foundational truth warrants. And then, in 1931, Kurt Gödel's astonishing work on incompleteness put a firm end to the idea that mathematical reason could be validly understood as entirely *a priori* anyway. See Nagel and Newman, *Gödel's Proof*. What we tend to call postmodernism is simply the recognition that modern ambitions to generate final first truths grounded in nothing other than sense and logic cannot, in fact, be obtained.
15 This line of philosophical (and philological) reasoning is much indebted to J.G. Hamann's brief yet devastating critique of Kant's critique of pure reason. See Johann Georg Hamann, "Metacritique on the Purism of Reason" [1784] in Hamann, *Writings on Philosophy and Language*, 205–218.
16 The field of study called the sociology of knowledge understands this very well. For a very accessible classic text in this field, see Berger and Luckmann, *The Social Construction of Reality*.
17 Epicurus, *The Art of Happiness*.
18 Kirk, Raven, and Schofield, *The Presocratic Philosophers*, 402–433.
19 Empiricus, *Outlines of Pyrrhonism*.
20 Lucretius, *On the Nature of the Universe*.
21 Gaukroger, *Objectivity*.
22 On the importance of distinctive forms of Christian theology to the decline of medieval natural magic and the rise of modern science, see Thomas, *Religion and the Decline of Magic*; Hooykaas, *Religion and the Rise of Modern Science*; Henry, *Knowledge Is Power*.
23 Lovelock, *Gaia*.

24 My very limited knowledge of what we Western modernists call Buddhism is indebted to this very helpful book by Burnett, *The Spirit of Buddhism*.
25 George Eliot's (Mary Ann Evans) English translations of David Strauss' *Das Leben Jesu, kritisch bearbeitet* as *The Life of Jesus, Critically Examined* in 1846 and her translation of Ludwig Feuerbach's *Das Wesen des Christentums* as *The Essence of Christianity* in 1854 were highly significant in the cultural crossing of a naturalistic materialist 'event horizon' in the Anglophone intelligentsia.
26 Nick Spencer has a close look at the complex shifts going on in theology and science in the nineteenth century in *Darwin and God* and *Atheists, the Origin of the Species*. Darwin is a very significant figure in this move towards naturalistic materialism, but – as Spencer points out – the relationship between a Christian theology of nature and Darwinian evolution is by no means necessarily defined by naturalistic materialism.

3 The Pre-Conditions of the West's Progressive Dominion Theology

Theology is like an iceberg: it is not the obvious bit that you can see that will sink your ship. It is what you cannot see, under the surface, which has orders of magnitude more inertial power than what you can see, that you should really be concerned about.

Hopefully, Chapter 2 gives some idea of the deep orienting momentum that collectively assumed first philosophy norms about nature have in any given culture. Dominant modern Western (and Westernised) members of the now truly global human civilisation are no different in this regard than the members of any other culture. We have unconscious collective first philosophy assumptions like everyone else. This chapter will now move from thinking about first philosophies and theologies of nature in general to the specific heritage of modern Western theology (as first philosophy) in shaping our approach to nature.

3.1 The Medieval Roots of Our Modern Western Theology of Nature

The Modern Western way of thinking about nature emerged gradually from a way of thinking that we now largely define ourselves against—medieval Christendom. Three astonishing medieval innovations in thinking were necessary to produce the modern world, and in this regard, the modern world remains deeply medieval in some of its most basic assumptions. These innovations can be loosely described as nominalism, voluntarism, and atomism. Modern conceptions of science and technology—and of the separation of religion from science and politics—could not have developed without these innovations. What I will describe as Progressive Dominion Theology could not have developed without these innovations. Let us look briefly at each one in turn.

3.2 Nominalism

Robert Pasnau is one of the most powerful scholars of medieval philosophy writing today and he is far from convinced that the term "nominalism" has anything like a clear or stable meaning. I do appreciate these scholarly concerns and would refer anyone interested in this remarkably fascinating area of intellectual history to Pasnau's beautiful text on medieval metaphysics.[1] But I am going to describe nominalism in simplified and generalised terms with the aim of allowing us to better understand our contemporary presuppositions, so my aim is different from following the detailed twists and turns of the astonishingly rich intellectual history that Pasnau explores so carefully.

The simplest way of outlining what nominalism has done to Western first philosophy assumptions is to broadly compare conceptions of realism from around the start of the high medieval era in the twelfth century with how we generally understand realism today. As Paul Tillich points out, it is helpful if you bear in mind that what we now think of as a realist outlook is "almost the exact opposite" of medieval realism.[2]

Peter Abelard was one of the great Western European intellectual giants of the twelfth century. What Abelard understood by the term *realism* looked a bit like this: those things that most early medieval thinkers took to be genuinely real were eternal and intellective—notably the medieval *transcendentalia*: Beauty, Goodness, Truth, and Unity. But to what we might call naïve early medieval realists, the intellective form in every mundane thing, such as the Ideal Table, which is partially expressed in every imperfect material table, also existed in some high intellective realm. The high realm of intellective *essence* is here thought of as genuinely real, and it is a realm of timeless, qualitative, divinely created truths. All form (the meaningful and purposive "shape" that any given being has) is gifted to existing things from the spiritual realm of essence (ideas). Here, physical and temporal *existence* is dependent on and derived from intellective and spiritual *essence*. Here, physical existence is also real, but transitory beings that exist in space and time are real in a derivative, not a primary sense. The physical participates in and is dependent on the spiritual to the medieval realist outlook.

Abelard was the first powerful thinker to deeply challenge the naïve early medieval realist outlook that Western Christendom had largely inherited from the late Classical era, via Augustine.[3] Abelard thought that universal essential truths were *flatus vocis* — mere words, just the breath of the voice. Another way of putting this is that universal and qualitative terms that don't refer to any particular concrete thing, are names (*nomina*) or ways of speaking, rather than real things in themselves. So *nominalism* is the stance that rejects early medieval essential realism and wants to understand reality only in terms of concrete existing things.

You can imagine that such a radical upending of the reality assumptions of the learned and religious class shaping the broader outlook of Western Christendom in the medieval era did not simply swing into play without any resistance. There is many a fascinating twist and turn in the rejections and revivals of older Platonist inflected Christian realisms that have been going on since Abelard, right up to this day. But in broad terms, two very influential fourteenth-century Franciscans—Duns Scotus and William of Ockham—largely carried the day in the high intellectual and theological sphere, such that the West's dominant intellectual culture has largely been nominalist since the late fourteenth century.

Three things are important to note here. Firstly, fourteenth-century nominalism happened in the high age of Aristotle. Secondly, it is post-Reformation nominalism that gives birth to modern natural philosophy, what we now call science. That is, the basic understanding of the nature of reality assumed in Western Europe's seventeenth-century rests on fourteenth-century nominalist assumptions. Thirdly, in significant regards, we are all nominalists now in the modern West, so the radical nature of this metaphysical revolution, let alone its impact on our first philosophy and theology of nature, is hardly noticed by us (let alone how medieval it is). Let us briefly unpack these three points.

3.2.1 Nominalism Happens in the High Age of Aristotle

Aristotle, like Plato, thought of "form" or "essence" as a real feature of reality. Yet, Aristotle differs from Plato in a significant regard as concerns form. To Plato, unchanging essence (Form) exists both prior to and independent of existing physical beings, even though all existing physical beings participate in Form in order to be beings of a particular kind. To Plato, matter must be united with the defining intelligible characteristics, qualities, and purposes of distinctive and differentiated physical beings (that is, form) for those beings to be recognisably a cat or a dog, a tree, or a rock, etc. Yet, in contrast with Plato, for Aristotle, Form does not exist independently of Matter in some purely intellective sphere (other, perhaps, than Aristotle's God) but Matter and Form require one another. Matter is the indispensable

Pre-Conditions of Progressive Dominion Theology 41

medium of Form to Aristotle. Which is to say that all essence had to be expressed in existence to Aristotle. There is no discretely transcendent realm of pure Form in Aristotle's thinking. In short, Aristotle is a more concrete and integral thinker than Plato when it comes to the relation between Matter and Form. Aristotle—who invented the Western intellectual discipline of biology—is also simply more interested in what we would now call science than was Plato.

With the recovery of Aristotle's corpus in the West in the thirteenth century, criticisms of what we would now call the idealism of Plato (which they then called realism) had a powerful ally in the new Aristotelian thinking. And as Richard Rubenstein points out, Aristotle's deep interest in understanding the natural world profoundly shaped medieval thinking and gave rise to modern science.[4] This last point is worth noting because the rejection of Aristotelian learning that is advanced by seventeenth-century innovators in natural philosophy is far more complex than we moderns often like to believe. We shall come back to this later.

3.2.2 Post-Reformation Nominalism Gives Birth to Modern Science

Between the fourteenth century and the seventeenth century, the most world-changing movement to happen to Western Christendom was the Reformation. This is not the place to explore the distinctive impact of Calvinist and Puritan thinking on the origins of modern science (and, of course, Catholics such as Galileo and Descartes are vital in the birth of modern science) but the point I wish to draw your attention to is that both the Catholics and Protestants—who drove what we now call the scientific revolution forward in the seventeenth century—were largely nominalists in their metaphysics. The implications of this for both first philosophy and theologies of nature are hard to overstate.

Modern science (and hence, Western modernity) was born nominalist. The concrete and the particular now define reality such that there is a more or less total inversion of the meaning of "realism" between the eleventh and the seventeenth centuries. By the seventeenth century, a realistic understanding of nature is built up by a carefully observed knowledge of particular concrete reality and its regularised appearances. Qualities, purposes, and intelligible essences in (and/or beyond) nature become obsolete. In important regards, Aristotle's careful rational observations of concrete natural beings is the first rocket booster that gets modern science out of the lower atmosphere, but it largely falls away in the seventeenth century. This falling away is worth understanding.

Aristotle observes four types of causes in nature: material, efficient, formal, and final. Formal causes are concerned with intelligible essence, and final causes are concerned with purpose. These two causes are simply dropped in the new natural philosophy of the seventeenth century, and the only type of causes we can recognise in nature now are material causation (what something is made of) and efficient causation (how one thing causes another thing, more or less mechanically). That is, our scientific realism has no conception of essence or purpose in nature. We have a simpler philosophy of nature than what Aristotle had, which is arguably simplistic as regards essence and purpose in nature. This does not make purpose and essence in nature go away; it just makes them invisible to what we think of as demonstrable scientific knowledge. But when purpose and essential value are no longer observable to our conception of a scientific understanding of natural reality, this promotes an instrumental and qualitatively un-constrained set of operational norms as regards what we do to nature with our knowledge and our technological power. Why not put huge quantities of sulphate aerosols into the atmosphere to "fix" global warming?

3.2.3 *Nominalism is Realism Now*

Nominalism in the Middle Ages was compatible with Aristotelian essentialism because individual existing beings that the nominalist could know were still thought of as concrete matter-and-form composites. But nominalism in the modern age is not compatible with qualitative or purposive or intellective essentialism. The significance of this is that qualities, purposes, and reason can no longer be seen by modern Western realism, *in nature*. By the late nineteenth century, we have come to think that *culture* puts all meanings, all qualities, all purposes, and all constructs of the human mind, within reductively naturalistic (i.e. purely material) reality. Here, Nature and Culture are in one sense riven from each other. In another sense, Culture is rendered an artifice and an epi-phenomenal fiction generated by blind and meaningless Nature. Here, nature is *really* without meaning or value or purpose. Culture is *really* just a constructed imaginative gloss on a meaningless reality. This has serious implications for how we understand both "bare" nature and our own now non-essential and culturally constructed human nature. But this sort of understanding is all under the surface as it were. We hardly even know where our distinctive cultural vision of reality comes from, and we hardly even know that it *is* a vision (as distinct from being "simply" real) and one that we *believe* by a long historical process of deep enculturation.

The Western development of the separate territories of science and religion over the past 300 years has allowed an objective realist vision of bare nature (bereft of all qualities and meanings) to go in one direction, and a constructed world of purely made-up human meanings and reasons to go in another direction. We think we still have values and meanings (and indeed we do still have them) but they are held strangely separate from natural reality, and placed in the subjective and constructed realm of personal belief freedoms. We have separated morality from nature and now we find there are no moral reasons why terribly damaging nature is actually wrong. We must find merely prudential reasons why a destructive relationship with the natural world is "bad". And where calculative prudential self-interest is the only measure of "good" and "bad", if I happened to be deeply invested in a structure of environmentally damaging power, then it is obviously prudentially self-interestedly "good" for me to keep on acting in my own commercial best interests, whatever damage that does to future generations. Our moral logic is not equipped with real essential values.

Who would have thought that Peter Abelard would still be having such a powerful influence on us after more than 800 years?

3.3 Voluntarism

In the declining years of the Western half of the Classical Roman Empire, Augustine (354–430 AD) preached and wrote in Latin. He was the bishop in the North African city of Hippo. Most of the great thinkers in the Christian world in its first flowering of high theology were Greek speakers who lived in the Eastern side of the twin-headed Imperial Roman eagle, centred around Constantinople rather than Rome. But back in the West, Boethius (477–524 AD) was the last Roman Christian thinker from late antiquity to be deeply educated in the high learning of Classical antiquity. The political unity of the Western Roman Imperium disintegrated in the sixth century, and whilst most of the philosophy, science, and high intellectual culture of antiquity was lost to the West at this time, they never lost Augustine. For this reason, the African bishop's thought towers over the monastic era of Celtic missions from the sixth to the twelfth centuries—the era in which Western Europe was converted to Christianity[5]—and Augustine's thought is deeply embedded in the medieval Christendom that flowered in the early and high Middle Ages.

In Augustine's thinking, the primary quality of God is love. As we are made in the image of God (in Christian theology), the attribute that most centrally defines our humanity is also love. But by a process that came to something of a head in the University of Paris in the late thirteenth century, a shift in theology made unhindered sovereignty—a totally free and all-powerful will—seem like the most defining feature of God, in whose image we are created. *Voluntas* (the Latin word for "will") thus gives its name to *voluntarism.*

Once again, Aristotle is intimately involved in this deep movement in the primary outlook of Western theology. Once again, fourteenth-century Franciscans prove deeply significant in laying the foundation for Western modernity.

When Aristotle's logic, science, ethics, and metaphysics were recovered in the West (by the latter half of the thirteenth century), the new intellectual institution of the Western University was coming into its own. The recovered ancient knowledge from Aristotle was far in advance of any other learning accessible to the thirteenth-century medieval scholar, so Aristotle was hungrily devoured and his integration with Christian theology was systematically pursued. The great thinker who baptised Aristotle into Western Christendom—the apex thinker of the University of Paris in his day—was Thomas Aquinas (1225–1274).

Aquinas maintains an Augustinian outlook on God and retains a Christian Neoplatonist metaphysics, and yet he was also a great lover of Aristotle and mastered the corpus of "The Philosopher" (Aristotle) with astonishing intellectual acumen. Other thinkers at this time embraced Aristotle so ardently as to at least appear to reject central creedal features of traditional Christianity. A Latin stance connected to the powerful Arabic Aristotelian scholar Ibn Rushd (Averroes) was so enamoured with the logic and order of the Aristotelian vision of the cosmos, that God became more of a rational and metaphysical principle than the living, feeling, historically acting God of the Christian scriptures. God as "thought thinking itself" and as the "unmoved first mover" (as in Aristotle's philosophical theology) would do nothing arbitrarily, and so the Deity's actions were—in effect—totally logically determined. This sort of necessitism was strongly opposed by ecclesial authorities in the thirteenth and fourteenth centuries.

To safeguard God's complete freedom from the fixed determinism of rational necessity, Franciscan thinkers like Duns Scotus and William of Ockham strongly emphasised that, for example, something is good because God decides it is good, such that God is not bound by necessity to do what we think is good.[6] In elevating God's sovereignty above a human understanding of universal and necessary reason, the divine will takes on a radically indeterminate and power-defined nature.

Pre-Conditions of Progressive Dominion Theology 45

There is a complex relationship between Augustine and the Franciscans here as love is seen by both as a function of free will, yet to Augustine a truly good human will has no freedom to choose to do what is evil, for such a "choice" would be a fall into the irrational bondage of sin. To Augustine, one only has a free will if one does what is reasonable and does not sin. To an Augustinian "intellectualist"[7] like Aquinas, love, will, and reason are fully integrated, which is to say that whilst reductive logic is unable to discern the qualitative reasons and free grace of love, love and goodness always define the deepest signatures of reason. To Scotus, however, the traditional understanding of the Augustinian intellectualist smacks too much of Greek necessitism.

God's complete sovereignty, as ungoverned by any external necessity, becomes a central feature of fourteenth-century Franciscan theology. God is totally free and all-powerful, and we are created in God's image, so we too are created for freedom and powerful creative action. By the early modern period rulership (sovereignty) over the physical world and all the creatures under heaven, and a creative will that can generate its own plans and intentions and follow them up (as distinct from being the necessary function of external causes and effects), were increasingly becoming central to the Western understanding of the meaning of being human.[8]

Have you ever wondered where the West's deep commitment to individual freedom and self-determination comes from? In large part, this cultural stance comes from medieval voluntarism. This outlook is now so embedded in our assumptions about what it means to be human that we don't even notice it as an assumption. An assumed identity right to rule over nature and to be self-determined in our actions is part of the deep theological foundations of the modern world. We uphold the principles of will and freedom in our deep assumptions of what it means for us to be human. And the field of action and ambition in which we exercise our will and freedom is the realm of nature.

Again, as with nominalism, voluntarism is a deep precondition for modernity. It is there well before the seventeenth century and is simply assumed as a defining feature of the human condition within Western modernity. After the Reformation, Calvinism is particularly associated with voluntarism. God's Sovereignty is central to Calvinist theology, and God's freedom (perhaps ironically) can in no manner be influenced by us, which has the effect of constricting us in a predestined necessitism that will determine our eternal destiny. Even so, as made in God's image, we too are creatures of will and desire freedom over our own sphere; thus, we are accountable for our actions but can never justify ourselves before God, as God's sphere of judgement is above our sphere of action and is not determined by us. The impact of a Calvinist understanding of the inscrutable and completely free will of God on both modern empiricism and the Protestant work ethic is a deep feature of the modern life-world.[9]

Voluntarism is a signature feature of Western modernity's culturally assumed first philosophy. Again, who would have thought that William of Ockham would still be having such a powerful influence on us after more than 600 years?

3.4 Pure Matter

The last precondition for a modern Western Progressive Dominion Theology that I want to draw your attention to is the idea of pure matter. This is a long and complex story, and again, we shall only touch on it briefly here in order to see how it shapes the first philosophy assumptions of Western modernity.

In the time of Socrates, a Greek philosopher by the name of Democritus (c460–370 BC) came up with the idea that reality was an entirely material affair, and that there were imperceptibly small and indivisible units of matter called atoms. Atoms—so Democritus postulated—were eternal and were in continuous random motion. Things appear to us to be hot and cold, wet and dry, and the cosmos might appear to be an intelligible ordered unity, but those meanings and beliefs are all conventional. The apparent world of our conventional meanings—according to Democritus—is actually an illusion. In reality, the only things that really exist are atoms, motion, and void.

The Platonist and Aristotelian trajectories in ancient philosophy strongly rejected ancient atomism. The idea that all apparent meaning and order in the cosmos is an illusion was treated by them as a denial of the validity and reality not only of common sense, but of reason, value, and intelligible meaning as well. Such a stance could not be taken seriously as—if true—it undermined the truth value of intelligent thought. Aristotle, however, made a sort of half peace with Democritus. Matter (though not the then unprovable theory of atoms) was taken as fundamental to existence by Aristotle (Aristotle's word for "matter" is "hyle"—the Greek word for "wood"), but Aristotle did not believe that there was ever such a thing as unformed matter. Aristotle's theory of matter is hylomorphic (a matter-and-form unity), which is to say that intelligible essence, qualitative attributes, logic, and innate purposes are inextricable from the material world to Aristotle.

Pre-Conditions of Progressive Dominion Theology 47

In the ancient world, atomic materialists were never more than a minority intellectual elite. The main forms of cultural belief were much more influenced by Plato, Aristotle, and the Stoics, as well as all manner of cults and what we would now call religions. Which is to say that ancient atomism never attained the status of being culturally mainstream in the ancient world. With the collapse of Classical Antiquity in the West in the sixth century AD, ancient atomism largely disappeared from the West. The belief outlooks on the nature of matter that continued in the West through the early Middle Ages were a blend of Christian Platonist outlooks and natural magic/pagan outlooks (this was before Aristotle was re-born into the West in the thirteenth century).

When the intellectual life of Western Europe was revitalised in the high Middle Ages, Aristotle is recovered at much the same time that the Western University is established as the pre-eminent intellectual institution of the West. A hylomorphic outlook on matter is more or less assumed between the thirteenth and seventeenth centuries. There are complex problems with the medieval theory of matter that develop that we will not get tangled in here. But what is truly amazing is that Democritus not only makes an astonishing comeback from the fifteenth century,[10] but also by the early twentieth century atomism achieves a cultural hegemony it had never had in the West before. This is a profound shift in the first philosophy assumptions of Western modernity, and it needs a little unpacking.

As standard twentieth-century accounts of the scientific revolution recount,[11] modern science rises in opposition to the Aristotelian-dominated university learning of the seventeenth century. This is true. Modern natural philosophy (what we now call science) was born on the fringes of the Western university. Apart from the fact that Aristotle got some pretty serious things wrong (inertia, projectile motion, and gravity, for example), the manner in which philosophy, science, and theology were richly integrated in late medieval natural philosophy made that field astonishingly complex. The pioneers of modern science grasped that our knowledge of the natural world could be considerably simplified if one reverted to experimental and mathematical description, as distinct from philosophical and theological explanation. Whilst Francis Bacon was setting up the outlook later embraced by the Royal Society of London to advance its experimental and practical approach to natural philosophy, the French humanist priest and man of letters Pierre Gassendi was advocating Democritean atomism as compatible with Christian faith and of great benefit in natural philosophy.

In very simplified terms, the Aristotelian natural philosophies of the sixteenth- and seventeenth-century universities were too complex to work well, aside from being significantly factually wrong in places, and the new empirical and practical sensibility of the times simply went around them, bringing the universities largely into line with the new trends in natural philosophy after the astonishing achievements of Isaac Newton. One of the things that got dropped as Aristotle went out of fashion was the hylomorphic understanding of matter. But there is a profound difference between seventeenth and eighteenth-century Epicurean atomism and ancient atomism, for early modern atomists were not materialists.

What became the characteristically modern and Western Christian separation of the spiritual from the material made it possible to think of the natural world in purely material terms, even whilst the material realm was itself a lower story of a two-tiered cosmos and was ultimately dependent on the divine upper story, at least for its origin. In this context, eighteenth-century Deism (which Hume attacks with such devastating power) thought of God as a divine clockmaker who constructed and then wound up the universe, only to depart "up-stairs" into "supernatural" Heaven; meanwhile, nature just ticks over by itself, "down-stairs". Ironically, it was the influence of eighteenth-century deism on German Liberal Protestant theology that gave rise to a reductive materialist atheism in the nineteenth century. By the nineteenth century, we no longer have an assumed dualist outlook where natural matter and super-natural spirit are separate realms, but we find the rise of reductive materialism fully at home with a practical and atomic notion of matter.

As science firmly marked itself off from religion in the late nineteenth century, the West developed, for the first time, a genuinely materialist atomism as a wide-spread scientific understanding of the real nature of the observable world. So, when it comes to observable material reality, our science doesn't think that anything other than atoms, motion, and void really exists. It becomes part of the practice of what we now think of as the scientific method to treat observable reality as if it is reductively material, inherently meaningless, purposeless, and without intellective essence (let alone continually held in being by divine creative grace).

Democritus reaches into our world from well over 2,000 years ago, profoundly shaping the way we see nature and human nature. We now largely see nature through the lens of reductively materialist atomic realism. Here, moral truths and religious beliefs are cultural illusions that have no true contact with natural reality. This has a profound impact on the way we understand nature, human nature, and the rights and wrongs of how we relate to the natural world.

3.5 Summary

We now have some appreciation of the nominalist and voluntarist preconditions for the West's modern scientific understanding of nature, and we now see how our culturally assumed philosophy of matter has swung in a profoundly Democritean direction during the course of Western modernity. Bearing this in mind, it is now time to look more explicitly at the distinctive Christian theological outlook that underpins the modern scientific age.

Notes

1 Pasnau, *Metaphysical Themes 1274–1671*.
2 Tillich, *A History of Christian Thought*, 142.
3 It is important to note that early medieval realism was, generally, of a pretty poor intellectual quality. The Christian Patristic Neoplatonist thinkers of late antiquity were far more subtle and sophisticated realists than anyone in the medieval West (bar John Scotus Eriugena in the ninth century) until Thomas Aquinas.
4 Rubenstein, *Aristotle's Children*.
5 See Cahill, *How the Irish Saved Civilization*.
6 Concerning Scotus: "…moral values derive from God; they do not restrain him." Cross, *Duns Scotus*, 90. Concerning Ockham: "God can directly do anything…" Keele, *Ockham Explained*, 108.
7 In simplified terms, an intellectualist is opposed to a voluntarist. That is "Reason" defines God's possible range of actions to an intellectualist, whereas "Will" defines God's possible range of actions to a voluntarist. An intellectualist's understanding of God is thus constrained by reason when God acts (in one way of looking at it) whereas to a voluntarist's understanding of God's all-powerful will is constrained by nothing other than its own free decision.
8 I would agree with Lynn White that a "dominion" outlook towards nature is present in the late Middle Ages. However, this is primarily a function of voluntarism in the fourteenth century and only becomes firmly linked to a dominion reading of the book of Genesis in the seventeenth century.
9 See Max Weber's fascinating study of the impact of Calvinism on the emergence of modern capitalism: *The Protestant Ethic and the Spirit of Capitalism* [1904]. See also Hooykaas, *Religion and the Rise of Modern Science*.
10 See Johnson and Wilson, "Lucretius and the History of Science." It is a fascinating feature of the late medieval era that explicitly anti-religious forms of ancient atomism, naturalistic materialism, and scepticism were read with keen interest by many a Christian scholar. Late medieval problems with Aristotle's philosophy of matter and ingenious means of

making Epicurean thought theologically safe for the Christian were significant contributors to this interest in alternative approaches to matter.
11 Steven Shapin notes that it is Alexandre Koyré who brings the term "the Scientific Revolution" into common use only in 1939. See Shapin, *The Scientific Revolution*, 2.

4 Progressive Dominion Theology

4.1 Two Types of Theology

In this text, the word "theology" has been used in two ways.

Firstly, it has been used to signify "first philosophy". Let us call this "Theology A". Here, theology is interested in the *grounds* of commonly held thought/belief/action commitments.

The second sense in which the word "theology" has been used in this text is in—to us—a religious sense. Let us call this "Theology B". Christian theology, for example, explores its divine first principles from within a set of doctrinal truth and normative practice commitments that are recognised as true and normative to other Christians who share those same commitments. The idea of "the Christian religion" can, in this sense, be distinguished from other "religions" that have different doctrinal and normative first truths. Now we can say, for example, that Islamic theology differs in some important regards from Christian theology in its primary doctrine and practice commitments.[1] Thus, we typically define "religious theology" in reference to the specific doctrinal, ritual, and ethical signatures of distinctive "religious" traditions.

Within what sociologists call the life-world of Western secular modernity, culturally assumed first philosophy (Theology A) and individually committed religious convictions (Theology B) run in and out of each other in an often fluid and un-noticed manner. Typically, first philosophy is more unconscious and collective than are modern understandings of religious theology. Religious theology, at least within the pluralist categories of modern Western secularism, is typically collective only in a freely chosen, enclave participation sense, and can even be individualistic in a way that floats free of any firm commitments to any particular religious community of belief and practice.

In seeking to understand the theological inertia of Western modernity in the direction of ecological degradation, we are going to need a solid appreciation of the now submerged Christian theological attitudes and assumptions that still shape the most ardent secularist and the most personally committed atheist who is yet a modern Western person.[2] For—however distressing and unwanted this may be—basic culturally shared attitudes to time, matter, will, personal freedom, nature, power, authority, and purpose in the West, are intimately derived from the Christian theological commitments that are buried in the cultural under-layers of Western modernity. These sub-strata were often explicitly forged by an intimate alliance between various forms of Christian theology and the new outlook of natural science, during the early modern period.

4.2 Christian Theology and Western Modernity

With some understanding of the medieval Christian roots of modernity, and bearing in mind how Theology A and B mingle in Western modernity, we are now able to explore the seventeenth-century theological foundations of Western scientific modernity. Early modern Christian outlooks on dominion over nature, on the meaning of humanity as in some manner distinct from nature, on freedom, and on progress, are of particular interest here. The process of radical secularisation that has gone on in Western modernity from the late nineteenth century to the present makes it hard for us to appreciate how Christian scientific modernity in the seventeenth and eighteenth centuries was. This factor makes understanding the relationship between theology and climate change in the present strangely hard for us to grasp. Yet the reality is, today's most irreligious and pragmatic business tycoon is successful and powerful in categories largely formed, and understood as good and valid, by the Christian theological categories of early modernity. We are not going to change the basic direction of our attitudes to power, freedom, and nature if we don't take a closer look at their underlying cultural justifications.

Let us explore the first philosophy foundations of scientific modernity. We will start with philosophical and practical concerns that look to be "merely" scientific, and move towards explicitly theological concerns from there.

4.3 The Apparent as the Real, Knowledge as Power

A significant transition from pre-modern to modern thinking concerns the erosion of the distinction between the apparent and the real. In the categories of Aristotelian philosophy that were dominant in the Western universities in the seventeenth century, this entails dropping the search for essential knowledge—the knowledge of how things really are—and focusing on how the apparent world appears to us, and how it works, instead. This did not initially entail the reduction of appearance to reality or knowledge to power, but that followed. Let us look quickly at how this initial shift happened.

4.3.1 Humility Concerning Essence

The gradual transition from the medieval natural philosophy of Aristotle to the modern age of science was characterised—as Robert Pasnau puts it—by the trading of depth for precision. As Galileo explains:

> ... in our speculations we either seek to penetrate the true and eternal essence of natural substances, or content ourselves with a knowledge of some of their properties.[3]

Endeavouring to leave high philosophical speculation to one side, Galileo pursued a humbler natural knowledge, grounded in the mathematical analysis of observable phenomena. This disaffection with the high philosophical ambition for essential insight—intimately connected with Aristotle and medieval theology in Galileo's day—and this turn to the precise observation of phenomena, comes of age with the scientific revolution's brightest star. Describing his own work, Sir Isaac Newton gives us this elegant picture of what his natural philosophy was all about:

> ... to derive two or three general principles of motion from phenomena, and afterwards to tell us how the properties and actions of all corporeal things follow from these manifest principles, would be a very great step in philosophy, though the causes of those principles were not yet discovered ...[4]

To Newton, every carefully observed natural phenomenon provides us with an opportunity to discover universal operational principles at work in nature, and this knowledge then opens up larger questions about why things are how they are. In this process of discovery, what we now call science keeps exploring nature and incrementally advances its understanding, but perhaps never comes to grasp the final reality of natural causes and intelligible essences. (Note: Newton, like us, had no idea what gravity actually *is*.) There is an explicit modesty in Newton's approach here, contrasted with what he sees as the empty speculative aims of Aristotelian natural philosophy that he so effectively displaced. Modern science, it seems, can know the manifest regular principles of accurately described natural behaviours, but lays no claim (yet) to essential and ultimate truth. What we now call Science seems to make lower-order truth claims than does Philosophy and Theology.

Even so, there is more to modern science than precise knowledge acquisition. Through our accumulated understanding of the properties of natural phenomena we gain a *useful* knowledge of the world. This interest in *working* knowledge is at the heart of our technological age. Francis Bacon explains that the guiding purpose of his vision of natural philosophy was not some hubristic ambition to understand metaphysical or theological mysteries, it was the humble pursuit of "human utility and power".[5] Modern science seeks to benefit humanity by gaining power over the forces, resources, and the dangerous caprices of nature. With this focus on useful knowledge, modern science does not simply trade metaphysical depth for phenomenological precision, it also trades meditative contemplation for active power. This does not mean that thinkers like Bacon, Galileo, and Newton had no theological or philosophical interests—far from it, as we shall see—but the precise knowledge of manifest and useful nature that they pursued was in direct conflict with the complex synthesis of medieval ecclesial theology and Aristotelian university philosophy that they effectively displaced.

In sweeping cultural terms, the functional isolation of a secularised Science from Theology and Philosophy, as this developed in the nineteenth century, entails certain life-world trade-offs: we gain observable precision as we lose metaphysical depth, we gain active power as we lose contemplative reverence. Indeed, to the fully matured scientific gaze, as Ivan Illich puts it, "observation replaces contemplation".[6] In this metricised and quantified observational mode, we isolate facts from meanings and values, liberating instrumental power from moral and theological constraint, and create a world of knowledge and power that is foreign to our actual human experience as valuing, thinking, embodied spiritual beings. This abstract, meaning-and-value-denuded epistemic gaze, we call objective knowledge. Perhaps strangely, we don't seem to mind the de-humanised, de-spiritualised knowledge we now have. This may have something to do with the deep-seated belief that we are ever-advancing through our science and technology. And considering what spectacular gains in knowledge and power the scientific

Progressive Dominion Theology 55

age has achieved, it could be that the trade-offs we made to get these gains are now largely invisible to us.[7]

4.3.2 The Passage from Scientific Humility to Scientistic Hubris

While this turn from essential knowledge to a logical analysis of apparent phenomena claims to be a humble approach, it is pretty clear that the great founders of modern science had a low opinion of any claim to essential knowledge from the outset. As science rose and culture evolved, the very idea of essential knowledge was increasingly treated with disdain, and the merely apparent world of universal causal necessities *became* reality.

Alasdair MacIntyre looks closely at the collapse of essential and moral knowledge that occurs at the same time as the reduction of reality to appearance and knowledge to power occurs in Western cultural history. This is a profound feature of the rise of Western scientific modernity.

MacIntyre pays close attention to the Enlightenment idea of empirical facts: this idea entails the isolation of experience from theory, the aim of observing the world in a presuppositionless manner, reducing reality to sensory experience "and nothing more".[8] Here, there is "nothing beyond my experience for me to compare my experience with, so the contrast between *seems to me* and *is in fact* [can] never be formulated".[9] The very idea of metaphysics, of the quest for a partial knowledge of how things really are, beyond the reductive empiricism of mere seeming, becomes increasingly philosophically disdained. The very idea of "ought" being a real truth category in the realm of ethics (if only "is" is apparent) becomes meaningless. Thus, humility concerning essence becomes pride in only dealing with appearance. Thus, recognition of a domain of theology and ethics that is outside of the humble scope of observable knowledge constructions of the apparent world in early modernity becomes subsumed *within* the domain of science in high modernity.[10] By the early twentieth century, the main carriers of Anglo-American philosophy had become positivist, linguistic, and pragmatic reductions to the domain of the apparent.

This has profound significance when it comes to attitudes to nature. If only the apparent is real, and if values and meanings are essentially subjective glosses that are discrete from objective reality, or only have reality when re-described in the instrumental and quantitative categories of empirical objectivity ("facts"), then there can be no objectively wrong way of dealing with nature, and there can be no sacred limits to human action within nature. We will return to this later. But there are explicitly Christian notions embedded in scientific modernity that we must now also understand.

4.4 Christian Theology and the Modern Scientific World View

As alluded to in the introduction to this book, many Conservative Western Christians seem to share an unstated common outlook with business-friendly political Conservatives who may or may not be Christians. This outlook is a broadly right-of-centre, free-market capitalist, profit-driven attitude to natural exploitation, and often entails climate change denial or inaction. This relationship usually looks contingent rather than integral because we like to think that personal religious conviction is discrete from the pragmatic worlds of business, politics, and finance. We are now at a point where we can unpack the ways in which these relationships are not at all contingent or only apparent, but are essential and integral.

4.4.1 The Rise of Science out of a New Secular Theology

Historians of modern science are intimately aware of the deep and mutually constructive ties between new forms of modern Christian theology and early modern science. Amos Funkenstein points out that

> A new and unique approach to matters divine, a secular theology of sorts, emerged in the sixteenth and seventeenth centuries to a short career. It was secular in that it was conceived by laymen for laymen. Galileo and Descartes, Leibniz and Newton, Hobbes and Vico were either not clergymen at all or did not acquire an advanced degree in divinity. They were not professional theologians, and yet they treated theological issues at length ... Never before or after were science, philosophy, and theology seen as almost one and the same occupation.[11]

As the actual history of the relationship makes abundantly clear, it is a new type of Christian theology that gave rise to modern science. Indeed, the defining features of

> the Scientific Revolution can be properly understood only against the backdrop of the theology which inspired and supported them.[12]

To be more precise,

> a distinctive feature of the Scientific Revolution is that, unlike other earlier scientific programmes and cultures, it is driven, often explicitly, by religious considerations: Christianity set the agenda for natural philosophy [i.e., modern science] in many respects and projected it forward in a way quite different from that of any other scientific culture.[13]

The need to harmonise the new experimental knowledge with the changing theological landscape of sixteenth and seventeenth-century European Christian thinking was the astonishingly innovative context in which modern science was born. Though our cultural memory has faded—and even though the attempt to replace that memory with a new origin story of a perpetual war between science and religion has occurred—scientific modernity is still deeply embedded in certain features of early modern European Christian theology. It is time to be more specific.

4.4.2 Protestants, the Bible, and Science

Protestant approaches to tradition, authority, and the "plain meaning" of things, are significant in the rise of modern science.[14]

The Reformation rises out of a "return to the sources" fascination with the Greek New Testament near the turn of the sixteenth century. The way ancient Greek manuscripts sometimes suggested different meanings to their authorised Latin translations provoked intense theological interest in Western Christendom at this time. As the printing press was up and running by the late fifteenth century, not only were Greek sources now widely available for scholars, but scholarly discussion by pamphlets and other publications gained a spread, a speed, and an inertia that facilitated the wild flight of new ideas in ways unknown until that time. Of course, complex wealth and power struggles between nascent rising nation-state powers and the self-preservation reflexes of the Church were by no means unimportant in stimulating impetus for church reform.[15]

When the Protestant Reformers broke with Rome, they rejected unquestioning submission to tradition-embedded ecclesial authority. Interpreting Scripture was a central issue here. If Scripture judged the Church, but only the Church could interpret Scripture, then the Church was likely to authorise only the interpretations that did not challenge its internal norms, power structures, doctrines, and common practices. The desire to free the interpretation of Scripture from the tight control of institutional power was a key driver of the Reformation.

In reading the Scriptures in ways that critiqued authorised ecclesial practices and doctrines, the Reformers maintained—as far as possible—a "plain meaning" interpretative philosophy. That is, the meaning of Scripture had to be reasonably demonstrable in its own terms to stand above institutional interpretive power. To be clear, Luther's understanding of the plain meaning of Scripture—its natural (non-allegorical) and historical (contextually situated) meaning—was not interpretively unnuanced. He was, after all, a highly trained bible scholar. Even so, from its outset, Protestants maintained a strong rhetorical opposition to the interpretive authority of tradition, and looked ascance at the construction of complex non-literal biblical meanings.

Peter Harrison's close examination of the role of Protestant scriptural interpretation in the rise of modern science is a very interesting study.[16] One thing that is clear is that this new questioning approach to traditional authority, and this new commitment to the plain meaning of things, moves from theology to early modern science. The 1662 motto of the Royal Society of London—*nullius in verba*—basically means "don't take anyone's word as authoritative" or "see for yourself". Harrison also points out that there is a strong Protestant influence that gives rise to a new and distinctly modern understanding of the "two books" (Nature and Scripture). A less traditional, more positivist understanding of both natural revelation and supernatural revelation was intimately interwoven in Protestant thinking in the early scientific era. In both forms of revelation, a plain meaning hermeneutic is upheld that actively disdains allegorical significations and metaphysical speculations.

By the seventeenth century, Protestant attitudes fed naturally into a disdain for any subservient submission to the academic authority of Aristotle in natural philosophy, and these attitudes upheld a distinctive type of common-sense empiricism adhered to by the group of English Puritans (suitably re-Anglicised after the Restoration) who became the Royal Society. Charles Webster and Reijer Hooykaas make particularly useful historical explorations into the distinctly Puritan theology that deeply facilitated the seventeenth-century rise of modern science in England.[17]

But it is not just an approach to authority and an interpretive positivism where theology shapes the new science and Western modernity. Christian theology shapes Western modernity through doctrine as well.

4.4.3 The Fall

Adam, so an Augustinian and Western reading of the Book of Genesis maintains, fell. It is hard to over-estimate what this Christian doctrine means for the culture of Western modernity and for shaping our tacit understandings of the relationships between humanity, nature, knowledge, and power.

You may think that nobody seriously believes in the Christian creation and fall myth anymore. Even if that were true, both conscious belief and conscious dis-belief in the fall—as expressions of "religious freedom"—are, in a very significant cultural sense, irrelevant. The fall of Adam is a Theology B doctrine in a modern religious sense, but it is also a deeply culturally embedded Theology A premise, in a world view sense. That is, whether you yourself are a young-earth six-day creationist, an allegorical creationist who integrates contemporary evolutionary and geological science with the Christian creation myth, a consumer pragmatist who is uncommitted to any firm religious or scientific truth claims, or a scientistic atheist who sees Genesis as complete baloney, the fall of Adam is deeply embedded in your unchosen world view reflexes as a modern Western person. Indeed, we would not have modern science as we now know it without this doctrine.[18]

In case you are unfamiliar with what the broad outlines of the early modern understanding of the fall is, here is a quick sketch.

> Adam and Eve were the first humans, placed in the enclosed garden (paradise) of Eden, by God. They were created in the image of God, who gave them authority to rule the earth. In their primeval innocence, they lived in perfect and harmonious communion with God. This communion was the source of peace with each other and with nature. They were naked and knew no shame. Originally, they had complete knowledge – by divinely illuminated insight – into the essential nature of all created things. Their thoughts and language were mediums of divine truth, and they named and ruled over all the earth using this true insight. This Edenic state was sinless, deathless, not marred by disease, hunger, thorns or hard labour, and entirely good.

> Two trees were at the centre of the garden: the Tree of Life and the Tree of the Knowledge of Good and Evil. Adam and Eve were given one prohibition by God; do not eat from the Tree of the Knowledge of Good and Evil. But they ate from that tree and so

fell, along with all creation, into a world marred by sin, suffering and death.

In their falling they lost perfect knowledge, they lost their harmonious rule over nature, nature itself became violent and fell into a different order that was a balance of good and evil (as distinct from original unmitigated goodness), and the man and woman were cast out of Eden. Outside of Eden humanity now had to learn how to live with sin, suffering, death, inequality between the sexes, conflict and murder, famine and hard labour. But they retained – in a tarnished manner – the image of God in their own natures. Fallen creation retained the tragic memory traces of the lost Eden such that good and evil were now mingled in nature. Adam and Eve were given a promise by God that the redemption of creation would be accomplished through the birth of a child. Eve conceived, founding the human race, in the hope that by her child the original harmonious order of creation would be restored.

This fall narrative profoundly shapes our Western understanding of the origin and nature of evil, defines our understanding of the nature of moral responsibility, and shapes our assumed collective sense that history has a redemptive direction. Our culture's assumption of linear time and our collective belief in progress are embedded in the fall narrative. The idea of dominion as belonging to the Creator, as gifted over nature to Humanity, and as now marred by the fall, is deeply embedded in Western political and legal norms. Our outlook on the limitations of natural knowledge after the fall, and the need to limit the political authority and power claims of our leaders (due to the sinful nature of fallen humanity) is deeply embedded in the doctrine of the fall. We will now look more closely at how the fall shapes distinctive features of Western modernity.

4.4.4 The Fall, Knowledge, and Scepticism

The notion of a fall—as Patristic Christian theologians of Late Antiquity realised—fits in neatly with an essentialist (i.e. Platonist) view of truth and knowledge. This patristic view of truth carries over, via various modifications, into the medieval era, and is still apparent at the start of the modern era. As we can only really see the distinctive features of what became the modern Western view of truth and knowledge when contrasted with other views, the theological history of how we now know truth is very rewarding in uncovering the background assumptions of our own era.

In Patristic Christian Platonism, every particular intelligible thing in the world has a true nature (a divine form), and that nature is only ever partially and unfoldingly actualised in the material and temporal world. For the world is subject to temporal flux, to spatial motion, and to change produced by causal contingency such that immediate sensory perception never amounts to a final knowledge of what something essentially is. Existence only partially reveals essence. Here, the cosmos is a creation, the Creator (as *Logos*) "speaks" essential intelligible meanings into the world, and yet *Logos* (the Divine Word, the source of all essential meaning) transcends the realm of spatiotemporal change. The primary essence of each being is firstly (and eternally) in the mind of God, and is then actualised in the *temporal* realm of existence, but in a way that retains its primary and essential dependence on God. To this outlook, things are only intelligible, only knowable by any mind at all (including animal minds), because all things have an unchanging intelligible essence that is partially expressed in space and time. This has striking implications for the way knowledge is understood from within a broadly Christian Platonist understanding of the fall.

We must look quite closely at the origins of modern thinking about true knowledge. In relation to the iceberg analogy at the start of Chapter 3, we are trying to look at deeply submerged features of our culture that have enormous inertia as taken-for-granted realities. This is going to be demanding, but it is very important if we are to see how what we now call scientific knowledge becomes embedded in a set of commitments that profoundly shape the way we treat nature. These commitments are born out of Western theologies of the fall.

Let us firstly unpack the essentialist understanding of truth that was still an embedded feature of the West in the seventeenth century. All of this will be unrecognisable to the three dominant truth theories of late modernity (correspondence, coherence, and pragmatic truth theories) but even so, the modern theories arise out of the complex interaction between the doctrine of the fall and the residually essentialist understanding of truth. And as truth itself is an essentialist notion, we remain more beholden to Platonist and Aristotelian conceptions of essential forms than we like to admit.[19] That is, because our three modern truth theories now explicitly reject essentialist conceptions of intelligible forms as features of reality, they unavoidably fail as accounts of truth, even if, in their own terms, they are workable—but not true—knowledge theories.[20]

To the essentialist Christian understanding of knowledge in the Augustinian tradition, the source of true knowledge is an uninhibited dialogue between the small "t" partial truth of the intellections of meaning that arise in our own minds, and the capital "T" essential Truth in the mind of God, as spoken into all of creation. The voice of truth in our mind, according to Augustine, is Christ, the *Logos* (Word) of God.[21] To Augustine, the capacity for intellection, and hence for meaningful thought, is inherently dependent on the active speech of the divine *Logos within* each human mind. Thinking truthfully is a dialogue between the Creator's Mind and the creature's mind about the reality that the Creator has created. Any internal truth discourse in the mind also points beyond creation to the splendour, love, and goodness of the Creator. Here, all created reality is continuously made by God *ex nihilo* (from nothing) and is totally dependent on God, the grounds of being, for both its intelligible essence and its concrete existence.

Seeing a pine tree, for example, is an act of communion between two minds about a being (the pine tree) that has an essence (its intelligible "pine-tree-ness" form) partially and particularly expressed in concrete existence, in the material world. Perceiving is an intelligible and meaningful act of a mind; it is not simply data reception. Perception comprehends the genus (essence) and particularised concreteness (existence) of the being perceived. But this act of the mind is a dialogue, it is a communicative enterprise between a truth receptive creature mind and the *Logos* of God "spoken" into the pine tree, giving it its partially realised intelligible form and concrete spatiotemporal material existence. Here, sinless fellowship with God is the first condition for an essential knowledge and wise understanding of created reality. Perception and logic need divinely illuminated essential and wise higher insight for us to experience the full glory of truth in the manner in which the Adam of paradise did.

If something goes drastically wrong in the dialogue of the thinking mind of the creature with the *Logos* speaking Mind of the Creator, such as sin,[22] then knowledge becomes unable to carry truth properly. This does not mean I cannot meaningfully perceive the pine tree if I am a sinner, but after the fall I perceive the tree in a manner that I do not often experience as a direct communication with God. God as the speaker is muffled in my mind, or only detectable in the tantalising traces of where God's prelapsarian intelligible immediacy used to be. After the fall the essential meaning of things becomes indistinct and I become less able to distinguish between eternal and essential reality and the uncertainties, flux, and contingencies embedded in the spatiotemporal medium of existence. The small "t" truths of my own mind become monologic poetic constructs locked within my own consciousness, rather than dialogic poetic responses to the transcendent Truth of the divine Speaker.

Originally, Western universities—such as in Paris, Oxford, Bologna, and Cambridge—were scholarly precincts set up as a doxological pedagogic enterprise by the Pope. The basic pedagogic purpose of the medieval university was to feed the hunger to know the divine light that infuses creation with its intelligible natures, and thus illuminate the devout soul. The diligent seeker of divine light (the seeker of God) was given wisdom as a reward for their efforts of love, yet after the fall, this labour required both divine grace and human effort. The light of wisdom (*sapientia*) was not simply essential knowledge, it was doxological knowledge, for true light is a vision enabled by God, and ultimately pointing to God, and the light of God gives all things their truth. Study, then, is a doxological act.

Explaining the pedagogic roots of the Western university, Ivan Illich notes that

> ... the relationship of things to God "who is light" must be understood. [The thirteenth century is] suffused by the idea that the world rests in God's hands, that it is contingent on Him. This means that at every instant everything derives its existence from his continued creative act. Things radiate by virtue of their constant dependence on this creative act. They are alight by the God-derived luminescence of their truth.[23]

As we shall unpack a little later, the original pedagogic vision of the Western university was being eroded by the sixteenth century, such that the natural light of *scientia* (knowledge) was being increasingly pursued autonomously from the essential illumination and doxological devotion of *sapientia* (divinely gifted wisdom). In response to that erosion, Oxford University adopted the motto *Dominus illuminatio mea*—"the Lord is my light" (from Psalm 27)—in their determination *not* to move with the times.[24]

By the seventeenth century, serious theological doubt about the claims of achieving essential insight via fallen reason was profoundly undermining the speculative Christianised metaphysics of Aristotle and the allegorical biblical interpretations of medieval Christendom. Particularly to Protestants, essential insight via any sort of scholastic pathway was increasingly seen as hopeless hubris. Reason and sense, as *natural* light, could not give essential truth, but they could be called upon for more humble and useful enterprises. Attainable accurate descriptions of natural properties, grounded in the secondary characteristics of the apparent world were fast displacing what was increasingly seen as unattainable essential truth.[25] It is no surprise that the recovery of the writings of the ancient sceptics at this time fell on receptive ears.

An important Neo-Platonist distinction that was radically reworked in the high middle ages, and then radically reworked again leading up to the seventeenth century, concerned the difference between uncreated intellective light (divine and mystical revelation) and created intellective light (natural perception and reason). The twin truth problems for fallen humanity are that direct intellectual illumination from the uncreated light of God is now a rare transcendently graced revelation, and the natural light of God, as mediating knowledge to the intellect through sense perception and careful reasoning, is corrupted by the fall and unreliable. Despite these obstacles, natural knowledge is not hopeless; attaining valid natural knowledge is a process that requires divine grace, hard truth-seeking work, and the assistance of the sacraments, the scriptures, and doctrines of the church to succeed. Considerable success was—with hard work and these aids—considered genuinely possible.

The medievals had a high, though subservient, opinion of natural reason. For it was clear to them that even though Aristotle was a pre-Christian pagan, he had gotten a lot right. Aquinas corrected his errors by reference to the Christian scriptures and Catholic theology such that natural philosophy and natural reasoning, drawing on the wisdom of the ancients, became enhanced by being bathed in the uncreated light of divine scriptural revelation. However, change was afoot. This change is hard for us to see because it happens in the period that Charles Schmitt calls

> the least studied period in the history of philosophy... the years separating Ockham and Oresme [fourteenth century] from Galileo and Descartes [seventeenth century] ... [This being a period of] about 300 years of the Aristotelian tradition, during which intellectual activity was more intense, the publications more numerous than at any time before or since.[26]

Significantly, the Reformation happens in this period of intense Aristotelian academic debate. Martin Luther famously called reason the "devil's whore" as a result of his disdain for the highly complex forms of Aristotelian philosophy and theology that dominated university life at that time. Luther was not opposed to reason, but he upheld Anselm's approach to faith seeking understanding, and thought the problem with reason occurred when this relationship was reversed such that reason sought faith. That is, Luther firmly maintained that natural human reason as a secondary truth discourse was fallen, and treated as a functionally *primary* truth discourse (perilous hubris!) natural reason displaced the light of divine revelation, and was then readily used by the devil to draw the faithful away from God. This outlook had much to do with Luther being an Augustinian monk and his intense conception of the fallen sinfulness of the natural state of Humanity, as well as his core Pauline commitment to redemptive grace depending entirely on the gift of God, rather than on any natural human capacity or effort.

The Reformation profoundly rocked the intellectual world of Western Europe. Enormous reactions and internal reforms within Roman Catholicism as well as political and social upheaval—and war—followed. By the time of Galileo and Descartes (both devout Roman Catholics), Catholics were expected to uphold a sharply non-Protestant outlook which usually involved affirming the authorised Aquinian-Aristotelian synthesis against the reason-despising Protestants, and affirming the ordered authority of the Church against the fractious divisions of the schismatic Protestants.

An interesting development happens in the seventeenth century as regards divine and natural light. Descartes wants to find ways of being certain about the natural light of reason. This is the Catholic origin point of modern rationalism. Even so, he is careful to distinguish the secondary nature of natural light from the primary nature of divine light. But the effect is to elevate natural reason as truth revealing within its proper sphere. Descartes does this as a Catholic who appreciates the difference between divine and natural light, but who also affirms the heritage of the synthesis of revelation with natural reason forged in the high middle ages. In this way Descartes can bat off a reviving ancient scepticism at the same time affirming the truth capacity of natural reason (that is, natural reason is fallen, but it still works in a limited way if you are careful with it). Descartes can, thus, be a good Catholic rejecting Protestant fideism and rejecting the Protestant radical rupture between natural human endeavours and supernatural grace.

English Protestants of the same period tended to go in the opposite direction but arrive at much the same point. This is the origin of modern empiricism, and it also is premised on the distinction between the divine light and natural light, but more defined by the epistemic dangers of fallen reason. Here, natural sense and reason are never true—somewhat agreeing with the Ancient Sceptics—but natural knowledge, if taken very strictly as secondary (not truth revealing), can be treated systematically so as to be *useful*. Faith becomes cordoned off from reason (the origins of the modern secular notion of religion). Faith is now a supernatural truth warrant and reason is a useful human tool rather than truth revealing. This is the origin of modern pragmatism as well. The natural philosopher can now deal only in sensory appearances and logically analysed cause and effect regularities in nature, without any ambition of attaining essential knowledge (true truth) thereby. But, pragmatic empiricism is a precise and carefully demonstrated form of secondary knowledge and useful power, so this also had the effect of elevating secondary natural light to the status of functionally certain proof. Natural reason could now be firmly separated out from the essentialist categories of theology. This stance disdains metaphysics—the idea that reason or science can penetrate essential truth—from the outset.

The above discussion is consistent with what we have already noted in MacIntyre's analysis of the Enlightenment translation of "seems" to "is" (the reduction of the real to the apparent) and the Baconian redefinition of knowledge as concerned with utility rather than essential truth. This development starts from the epistemic implications of the doctrine of the fall, and this development comes to define modern thinking in such a way as to eventually occlude essential truth from the field of reasoned knowledge; natural *scientia* (knowledge) becomes entirely autonomous from divine *sapientia* (wisdom).

Theology itself completely changes character in this process. Modern Protestant theology, as an academic discipline, becomes the application of natural logic and natural science to the texts that contain supernatural revelation. This gives rise to the modern and discretely religious intellectual discourse of Systematic Theology. Thus, by the late nineteenth century, the discretely religious truths of divine revelation become removed from the arena of public knowledge and safely quarantined in the seminary. At the same time, entirely natural processes of knowledge are being applied to understanding divine revelation. Revelation itself is not understood as essential insight, but as an entirely natural product. Once theology is safely contained in a religious though increasingly naturalistically understood domain, the rest of natural knowledge can now escape its subservience to divine wisdom. Natural knowledge (natural light) can now assert its autonomy from its old master, transcendent and essential knowledge (divine light).

Historically back-peddling a little, the wondrous success of the new natural philosophy —particularly after Newton—functionally justified a logically and empirically "bare" natural knowledge,[27] and a purely instrumental technological power as concerns material reality. Thus do bare knowledge and instrumental power—where both are disconnected from any transcendently framed and essential moral realism—become the demonstrable truth categories of the modern age. From the mid-eighteenth century, theology (as Deism), and ethics (as qualitative truths reframed as logical necessities or pleasure and pain quanta calculations) start to come under the umbrella of an entirely natural-light-defined Reason and Science. God and the Good are now defined (and then abandoned) in the purely natural knowledge categories of empirically demonstrable facts and logically necessary relations. The very idea of essential truth gets lost in this process, and the value-neutral, objective, merely logical and purely factual become the carriers of truth. This all comes to a head in the late nineteenth century when Science displaces Theology/Philosophy as the West's first truth discourse. This majestic glacial movement in the very meaning of Christian theology and natural knowledge—a movement from the fourteenth to the nineteenth centuries—takes modern Western Christians by surprise. Its end result—the separation of knowledge from revelation, and the public truth priority of reductively naturalistic knowledge over religious revelation—produces an enormous cultural upheaval in Western modernity.

With regards to environmental degradation, we no longer have any essential truth categories for the inherent value of anything (such as an endangered species), and we no longer have truth categories that locate moral imperatives in transcendent reality (moral realism of any sort is now a hopelessly nostalgic sentimentalism). Nature has become an object to us, morality has become our own creation, purpose and inherent meaning are outside of reason, and we are enamoured with instrumental power, value-neutral facts, and constructed meanings. The very idea of the sacred or any collective accountability to a Creator has escaped our public knowledge discourse and our secular power structures entirely. This process was intimately connected to the way our early modern Christian theology of the fall shaped our approach to natural and uncreated light.

No other culture in the history of humanity has "succeeded" in breaking away from the sacred as thoroughly as twentieth-century Western modernity has. With this breaking away comes an absence of restraint regarding what everyone else in the history of humanity has respected: sacred taboos. We think of this as progress, as enlightenment, as evidence of advancement—but it may kill the planet as we know it, and us with it.

4.4.5 Eschatology, Dominion, and Progress

As Lynn White outlined, a traditional Western Christian understanding of history has a particular shape and direction. Here, creation and redemption history happens in four acts. In the First Act, there is an entirely good creation. In the Second Act, there is the fall, and the cosmos becomes qualitatively ambivalent. Reality as we now experience it is defined by a fallen natural order in which good and evil coexist in a somewhat touchy stasis. In this part of the narrative—between fall and redemption—tyranny and vice may succeed for a while, but there are moral and spiritual realities that make human wickedness self-destructive. On the other hand, a good social order tends naturally towards corruption. This Western vision of a morally troubled, decaying and also self-correcting, and divinely overshadowed universe, is beautifully articulated in Shakespeare. Even so, this complex fallen stasis is not forever. The major divine intervention towards redeeming humanity and the entire creation is the incarnation of Jesus Christ, followed by his passion, resurrection, and ascension. This initiates the Third Act—the age of the Church as situated within fallen nature—leading to the Fourth Act in the return of Christ at the Day of Judgement (the eschaton). The lead-up to the Day of Judgement is going to be catastrophic, but it is the gateway to the healing of both humanity and nature from the blights of sin, sickness, the devil, and death. The Last Day will come, but when times get particularly bad Christians are instructed to "look up, for your redemption draws nigh."[28]

In times of great insecurity and upheaval, a sense that the world is coming to an end is a recurring feature of Western civilisation. When this happens, Western culture's deep embedding in the cosmic biblical narrative usually unleashes both eschatological fear and nervous anticipation. Significantly, the political and religious turbulence and the violence of the seventeenth century strongly promoted end-of-the-world sentiment during the crucial formative period of modern science.[29] What is noteworthy here is the way Francis Bacon—arguably the most influential figure in the rise of the new natural philosophy—uses his own distinctive eschatological outlook to justify the project of recovering dominion over nature. This, again, is intimately connected to the Christian idea of the fall. Bacon explains:

By the Fall, man fell from both his state of innocence and from his dominion over creation. But even in this life both of these losses can be made good; the former by religion and faith, the latter by arts and sciences.[30]

Redemptive history moves towards the healing of the fall. Bacon was a man of deep religious devotion[31] and he saw his new practical natural philosophy as integral with three highly significant redemptive missions: to regain Man's rightful dominion over nature, to improve the conditions of human life through that dominion, and to hurry up the end times and usher in the Kingdom of God by advancing the increase of knowledge.

Significantly, the illustration on the cover of the 1620 edition of Bacon's *Great Instauration* has the biblical prophecy from Daniel 12:4 written in Latin as a motto underneath a picture of a ship venturing out into the Atlantic Ocean past the Pillars of Hercules. The picture refers to the discovery of the Americas and the then recent reality of global travel and conquest. The motto runs, *Multi pertransibunt et augebitur scientia*: "many will go to and fro, and knowledge will be increased."

Bacon explains:

> What else can the prophet mean ... in speaking about the last times ... does he not imply that the passing to and fro or perambulation of the round earth and the increase or multiplication of science were destined to the same age and century?[32]

Bacon was the opposite of a closeted academic; he was a man of action, of influence, and of political vision. He firmly believed that the powers of the nascent modern state should be used for the advancement of learning, for discovery, for war, for industry, for commerce, and for the re-creation of human society governed by a priest-class of scientists. Almost 400 years after its original publication, Bacon's *New Atlantis* remains a gripping and inspiring utopian pamphlet.

Modern Western Humanity, created in the image of Francis Bacon, is humanity with a progressive scientific and technological mission. This is a mission of uncovering all the secrets of nature and harnessing her resources for the utility and improvement of the physical conditions of human life.[33] This is a vision of progress towards some ideal and perfect human reality (the Kingdom of God on earth, no less). This is a vision where advances in science and technology are the markers of advances in the human race. This is a voluntarist vision where ever-greater freedom will be available as ever-greater powers are discovered by science and harnessed by technology. It is a vision of power over nature through science and technology; science and technology enable us to reclaim our rightful Adamic dominion over the earth and all its creatures.

In the past century and a half, the religious drivers that shaped the original trajectory of the scientific age and propelled it forward have become submerged under the waters of time. But those drivers have not gone away; they have simply presented themselves to us in secularised forms.[34] Now we have the problem that we remain a people of cosmic mission, progressing courageously towards some perfection of the human race, but we have lost the transcendent and religious horizon that gave us that mission in the first place.[35] We have enormous instrumental power, but the very idea of wisdom and of qualitative and essential truths are no longer culturally available to guide us in the good use of our powers. Yet the inertia for progress, the infatuation with technological power, the uninhibited commitment to our own freedoms expressed through power, and our sense of right to rulership over the resources, creatures, and less powerful fellow humans of the earth, is an indelible signature of Western modernity. This is a signature—whether we are personally religious or not—that is inherently theological in its nature.

When the End of Time was upon Western culture in the sixteenth and seventeenth centuries, the response and path forward was found by science, technology, government, and trade. That—at least for the dominant classes of the European West—really worked (or so we believed). The idea that more science, more technology, more government, more trade will solve any crisis we face is now profoundly imprinted on the assumptions of our collective consciousness. The notion that we might slow economic growth—let alone stall it, or wind it back to make economic activity genuinely environmentally sustainable—is now politically unthinkable. Our leaders must triumph for us in ways that *advance*, rather than retreat. The idea that if we should not use some known and profitable technological power, seems almost immoral to us. For instrumental power, itself, is a deeply embedded cultural good to us.

In the deep underlying cultural imaginary of modern Western belief, the closer to the eschaton that we get, the more determined the logic of dominion and power becomes, because this has been our pathway to advance and victory for the past 400 years. Turning that sort of cultural iceberg around simply doesn't seem possible. For that reason, a "realist" stance wants to try and harness the existing forces of technological advance, economic growth, and government control to find ways forward, rather than fundamentally re-considering the very aims and direction of our cultural inertia. The ways in which the logic of eschaton, dominion, and progress are hard-wired into our cultural reflexes, well beneath the surface of our conscious choices, are a function of both the Christian Theology B narrative embedding of our culture, and of the Theology A first commitments of our civilization. If we want to change our ways and not simply degrade the biosphere to the point of collapse, we are going to need to reform our theologies.

4.5 Progressive Dominion Theology is a Primary Cause of Climate Change

What I have called Progressive Dominion Theology (PDT) is Western Modernity's integrated Theology A and Theology B package. That is, Western Modernity's Theology A is deeply shaped by a linear goal-defined conception of time, voluntarist assumptions about will/freedom, nominalist assumptions about meaning, non-essentialist understandings of knowledge, philosophically atomist assumptions about matter, and the reduction of moral and spiritual truth to cultural constructions of power and material necessity. Integral with these first philosophy signatures is the historical role that Western Modernity's Theology B has played. By now, Christian eschatology has been secularised into progress, early modern Christian conceptions of dominion have been secularised into an infatuation with instrumental power, and Christian conceptions of creation as provided to us by God are secularised as a world of natural resources to plunder. This is Progressive because it is always driving forward; we must advance in science and technology, we must have economic growth, we must overcome all oppositions to our freedom and desires. This is a Dominion theology because human rule *over* nature is simply assumed to be our basic right, a primal marker of our identity as humans. This is a Theology because it is a tacit cultural first philosophy orienting our collectively assumed world view.

Undoubtedly, an enormous array of factors are significant in getting us to our present crisis of anthropogenic climate change. For example, the formation of the Bank of England in 1695 invented modern state-backed credit which made modern capitalism possible. This matured in the 1880s with the creation of utility corporations and large-scale banking in the United States, unleashing unprecedented corporate investment coupled with unprecedented technological capabilities, such that the resource exploitation of the world has accelerated out of all historically imaginable proportion in the twentieth century.[36] Yet, underneath all these factors, PDT is the underlying value, meaning, and identity narrative—the cultural mythos—shaping Western modernity when it comes to our relation to nature. PDT profoundly defines our collective assumptions about knowledge, power, and nature. This is not something we choose or reject, it is the framework of common meaning into which we are born, like our language.

PDT is the primary culturally causal factor in our failure to deviate from the path of anthropogenic climate change. A "born to rule" technologically empowered consumer culture with an unconscious and secularised eschatological goal gives us unbending confidence that any improvement in instrumental power over nature (via science) is progressing us towards utopia. Ironically, it is science that is telling us we are going too fast, exploiting nature too hard, and irrecoverably damaging the balances upon which our own material wellbeing and, even more drastically, the wellbeing of future generations depends.[37] However, the theology that gave us modern science is simply not listening to *that* scientific voice. Powerful commercial and political forces in our now global civilisation continue to culturally believe that we are entitled to have power over nature on our terms, come what may.

The Christian theological origins of PDT remain deeply culturally embedded in the secular West. Yet, even as the West has become increasingly post-Christian, the first philosophy assumptions that have made Western modernists masters over nature will continue to shape us, though without any form of religious or transcendently accountable restraint.[38] This is a very dangerous mix. Linear and project-defined notions of time and vocation remain just as active in our more post-Christian culture as they were in the "Christian" colonial era. Whilst the colonial West certainly had a born to rule mentality then, the West's public consciousness no longer has any sense of accountability to God. We no longer regard creation as firstly belonging to God, to whom we must give an account. The horizon of public morality is now situated entirely within what Charles Taylor calls the immanent frame.[39] This means there is no transcendent right and wrong, and all values are cultural constructs now. That is, the scientific facts may tell the Adani mining company and the Australian government that using coal as a power source will increase global climate instability, but there is no objective moral meaning to those facts. The company "values" its profit-making powers and the Australian government "values" the royalties it receives from the mining sector. So profit and power become the defining moralities prioritising coal mining over addressing global warming. Without any public horizon of theological accountability, without any objective conception of

Progressive Dominion Theology 73

morality, we seem to have produced an entirely pragmatic public morality and an entirely materialist/hedonist voter spirituality.

Lynn White remains largely right. Western theology, at a very deep cultural level (Theology A) as well as a strongly modern form of Christian theology (Theology B), is the most basic life-world settings driving us over the climate catastrophe cliff. Yet, White thought Christian theology was also the best source of a world-saving ecological transformation for Western modernity as well. The rest of this book will look at what is going on in Christian theology both in ways that seek the change White hoped for, and that hinder that change. Understanding that dynamic may yet be very significant in any strategy that seeks to tap the theological drivers of our anthropogenic climate change.

Let us, then, have a look at what is going on in Christian eco-theology.

Notes

1 It is equally true that Islam and Christianity share key common doctrinal commitments, such as a monotheist and transcendent doctrine of God.
2 See Gregory, *The Unintended Reformation.*
3 Galileo, third letter on sunspots, 1612. As cited in Pasnau, *After Certainty*, 14.
4 Newton, *Opticks* (Latin edition of 1706), query 31, 401–402. As cited in Pasnau, *After Certainty*, 15.
5 Bacon, *New Atlantis and The Great Instauration*, 16.
6 Illich, "Guarding the Eye in the Age of Show," 18, note 64.
7 Or – if we are unreconstructed modernists – we do not see them as trade-offs but as pure and simple gains. Here metaphysical speculation is inherently meaningless, and sacred taboos concerning the use and non-use of nature are inherently irrational and misguided.
8 MacIntyre, *After Virtue*, 80.
9 MacIntyre, *After Virtue*, 80.
10 Utilitarian ethics reconfigures pre-modern moral categories so that they are amenable to observable and measurable sensations and behaviours (the realm of appearance). The rejection of any miraculous claim as true in the modern scientific age as applied to biblical textual analysis and theology situates divinity within the categories of scientific credibility (the realm of appearance).
11 Funkenstein, *Theology and the Scientific Imagination*, 3.
12 Henry, "Religion and the Scientific Revolution" in Peter Harrison (ed.) *The Cambridge Companion to Science and Religion*, 42.
13 Gaukroger, *The Emergence of a Scientific Culture*, 3.
14 This is not to say that Roman Catholics – think of Galileo, Descartes, Gassendi, for instance – are not equally central to the rise of modern science.

74 *Theology and Climate Change*

15 Cavanaugh, *The Myth of Religious Violence*.
16 Harrison, *The Bible, Protestantism, and the Rise of Natural Science*.
17 Webster, *The Great Instauration*; Hooykaas, *Religion and the Rise of Modern Science*.
18 See Harrison, *The Fall and Man and the Foundations of Science*.
19 The idea that what something really is is distinct from simply how it appears to be, or how it can be manipulated to look like one thing or another thing – the idea of truth – is an essentialist idea.
20 Pasnau, *After Certainty*, 1 notes that "the very term 'epistemology' goes back only to the middle of the nineteenth century." Our three modern "truth" theories arise within this new field of epistemology. Something of a philosophical sleight of hand has occurred in this nineteenth-century development such that what has really happened is that "epistemology" veils the attempt to dispense with any essentialist notion of truth, and rebrands non-essentialist "knowledge" as truth.
21 See *The Teacher* in Augustine, *Against the Academicians and The Teacher*.
22 "Sin," in Christian theology, is not primarily about performing an immoral or divinely forbidden act; it is an inherent failure to realise the proper created purpose of being human. Sin is, in Christian theology, an ontological category before it is a moral category. See Yannaras, *The Freedom of Morality*, 13–48. To Christian theology, sin and faith are unavoidably entailed in any attempt to apprehend (and, more importantly, be apprehended by) truth. See Kierkegaard, *Concluding Unscientific Postscript*.
23 Illich, "Guarding the Eye in the Age of Show," 16–17.
24 Change comes slowly to medieval institutions like Oxford and Cambridge. To this day you can still participate in a superb Latin evensong at St John's in Cambridge. You can see why Thomas Huxley and John Tyndall had to set up the X Club to carefully undermine the influence of the church over the established universities in the late nineteenth century.
25 It is also the case that Francis Bacon redefines – in some contexts – "Form," to mean "basic material structure, the way in which [any given thing's] constituent material parts are disposed" and also as a word for natural laws of operation, or simply a description of basic properties (Gaukroger, *Francis Bacon*, 140). That is, Bacon redefines "form" in reductively materialist and atomistic terms. In other contexts Bacon uses "eternal and immutable Forms" to distinguish metaphysics (which he has no time for) from physics and mechanics (Gaukroger, *Francis Bacon*, note 18, 141). Bacon had a complex interest in Lucretius and Democritus – the ancient atomists – so when he uses "form" positively, he is not using it in an essentialist manner. The rejection of essentialist thinking, whatever terminology he uses, is characteristic of Bacon's thought and of modern scientific thinking with roots in the Baconian method.
26 Schmitt, *Aristotle and the Renaissance*, 3.
27 See Agamben, *Homo Sacer*.
28 Luke 21:28.
29 Two outstanding texts that show the direct relationship between sixteenth- and seventeenth-century apocalyptic thinking and early science are: Henry, *Knowledge Is Power*; Webster, *Paracelsus*.
30 As cited in Gaukroger, *Francis Bacon*, 76. From *Nov. Org.* II.52.

Progressive Dominion Theology 75

31 Henry, *Knowledge Is Power*, 108. Henry points out that Bacon was from a deeply religious family and an inescapably religious era. The idea that the founders of science must have somehow been secularists and at least closet atheists is historically false. Henry notes (*Knowledge Is Power*, 20–21): "For Bacon and his contemporaries, God and religion were so pervasive in social, political and intellectual life that, to a large extend, systematic disbelief was practically impossible ... in Bacon's day atheism was barely comprehensible in a world that was thoroughly suffused with religion."
32 Henry, *Knowledge Is Power*, 23. This is from Bacon's 1608 pamphlet *Refutation of Philosophies*.
33 Carolyn Merchant, a key figure in eco-feminism, has interesting things to say about a masculine approach to the violent and instrumental interrogation of feminine nature in Western modernity. Bacon is here read within the Western misogynist tradition. See Merchant, *Death of Nature*.
34 See Cavanaugh, *Migration of the Holy* for fascinating studies of this sort of transition.
35 Nietzsche, *The Gay Science*, section 125, "The Madman." Nietzsche grasps very clearly that the cultural loss of God wipes away the horizon of our own meaning as creatures and wipes away the framework of divinely given purpose.
36 See Goodchild, *Theology of Money*; Galbraith, *The New Industrial State*; McNeil and Engelke, *The Great Acceleration*.
37 Houghton, *Global Warming*.
38 See Ellul, *The Technological Society*; Virilio, *The Great Accelerator*. The commitment to speed and efficiency, without any purpose other than increased speed and efficiency (and increased profit) points to the self-destructive futility of our secularized eschatology.
39 Taylor, *A Secular Age*.

Part II
Towards Christian Theological Solutions to the Global Ecological Crisis

5 Contemporary Christian Theologies of Nature and Climate Change
Roman Catholic, Celtic, Orthodox, Indigenous, and Mainline

Invoking again an iceberg analogy, the submerged cultural inertia of our life-world's tacit theology is more powerful than we readily acknowledge, precisely because we don't even know it is there. But there are some strikingly unusual features of the *visible* bit of the Christian theology iceberg under the conditions of modern liberal secularism.

Firstly, in many areas of the Christian world, theologies of nature have changed dramatically since early modernity, and particularly over the past 50 years. But secondly, in virtue of Christian theology itself now residing within the hermetically sealed personal freedom realm of religion, the visible bit of Christian theology has an astonishing *in*visibility in the public gaze of Western modernity. The religious nature of theology itself now means that Christian theology seems powerless to speak to any errors within our culture's theology.[1] This is a tragic shame because what is most striking about many trajectories in contemporary Christian theologies of nature is how strongly opposed they are to PDT.

Let us start with Pope Francis' 2015 Encyclical *Laudato Si'*.

5.1 A Contemporary Roman Catholic Theology of Nature – *Laudato Si'*

Pope Francis starts his encyclical about our common home quoting Francis of Assisi. The Canticle of the Creatures opens with

> Praise to you, my Lord, through our Sister, Mother Earth, who sustain and governs us.[2]

Most deliberately, Pope Francis starts from a stance that is an inversion of the power and use focused outlook of PDT. In this canticle we are related to and dependent on the earth, and our relationship with her is one of love and submission, not of controlling voluntarist use.

The theology of nature outlined here is, from the very start, anti-voluntarist,[3] anti-pragmatist,[4] upholds essential meaning,[5] and it is situated in a cosmology that is overshadowed by transcendence.[6] The Pope's stance denounces an "irrational confidence in progress"[7] and boldly owns the fact that "it is human causes that produce and aggravate climate change".[8] The Pope also recognises that it is the poor and vulnerable who suffer most directly from the environmental degradation produced by climate change and the rising global scarcity of basic resources such as clean drinking water.[9] Climate change is not simply an ecological problem; it is an international moral problem of the first order. A plunder logic drives our economic, commercial, and production norms, and this the Pope finds to be incompatible with a Christian creation theology where all creatures are intrinsically valuable in themselves.[10]

At a basic cultural level (Theology A), the Pope finds that we lack what is needed to confront the crisis we have produced.[11] At a Theology B level, we have lost an awareness that hubris, sin, and merely dominating power are always problems within fallen human nature that must be firmly resisted. Without recognising sin and without acknowledging our moral failure, we and the earth itself become subjected to a destructive "tyrannical anthropocentrism"[12] which arises out of a distorted conception of dominion,[13] and which is ultimately a function of idolatry.[14]

The above is a highly distilled summary of the first two chapters of *Laudato Si'*. I could go on with a similar summary of the remaining four chapters, but the above is adequate. What this encyclical makes abundantly clear is that the Pope—the highest teaching authority for 1.3 billion Roman Catholics worldwide—completely rejects the idea that modern PDT is theologically valid for a Christian. The reasons for his rejection of this stance are not spelt out in much depth in *Laudato Si'*, as the focus of the encyclical is practical and moral. This encyclical is part of the social teaching of the Catholic Church so it is not written for the academic specialist but for the practising Christian, and for all people with a deeply concerned interest in the need to nurture and protect our common home.

5.1.1 Laudato Si' and the Modern World

A significant thinker present in the background of *Laudato Si'* is the German theologian, Romano Guardini (1885–1968). The Pope draws on Guardini's *The End of the Modern World* (1950) at crucial moments in his argument about the human roots of our contemporary ecological crisis.[15] Significantly, Guardini is an important figure in the serious attempt to recover the deeper wisdom traditions of the Church, which entails a sophisticated intellectual return to medieval and patristic sources in first-order thinking.

Key figures in the past century in Roman Catholicism have uncovered an understanding of Western modernity that is not defined by the modern myth of progress. In the context of *Laudato Si'*, Guardini stands for a Catholic theological understanding where there is no reason why a modern stance is automatically superior to a medieval or classical understanding of the meaning and purpose of nature. Indeed, scientific modernity offers *no* account of the meaning and purpose of nature, but that does not necessarily imply that nature is truly bereft of meaning or purpose. This critical ability to be within modernity, yet view its strengths and weaknesses in the light of pre-modern Western theology and philosophy, has influenced the Roman Catholic Church at its highest level, particularly during the last two pontificates. This is often called the *Resourcement* movement—a return to the pre-modern sources of Christian theology.[16]

The French Jesuit Henri de Lubac (1896–1991) is also a key figure in releasing Catholic theology from the ironically modern "manual theology" that dominated its intellectual circles from the seventeenth to the twentieth centuries. Tight divisions between nature and grace and a rationalist way of doing both natural theology (as theology not drawing on grace) and revealed theology (as theology not drawing on nature) were profoundly challenged by de Lubac.[17] A key component of those separations was the rise of the modern understanding of religion. Here, religion is a purely spiritual matter, discrete from natural reality. That nature has been able to be mastered and harnessed outside of religious constraints and outside of any overarching theology of creation has been a crucial feature of the rise of Western modernity, and of the eventual separation of science from religion in the late nineteenth century. The twentieth century was born as nature and grace were being fully disengaged in the intellectual culture of Western modernity. This, of course, was not only happening inside the religious realm, but was, by the early twentieth century, driven by powerful atheist and agnostic thinkers, particularly of a Progressive social reforming bent (that is, advocates of a secularised eschatological sense of purpose in history, usually deeply indebted to Hegel).

Getting nature and grace to re-engage now that they have been culturally and politically uncoupled, and now that we have had the 1960s revolt against Christian customs, scriptures, and morals, is perhaps impossible. Getting theology out of a discretely religious box (i.e. the personal and "supernatural" box) and into the public and secular realms of knowledge, law, action, and power may now also be impossible without (perilously?) converting Christianity into a special-interest lobby group. This is why the Vatican-endorsed Christian theology of nature can be totally opposed to PDT, and yet PDT continues to shape the mainstream of power and practice in democratic, liberal, and secular Western modernity.

De Lubac and Guardini seem to suggest that the only way for Catholic theology to re-engage the natural world in which we actually live, with the political structures under which we now live, is to—in some manner—denounce modern religion. Pope Francis also appreciates something of this, and yet, as the Pope, he is very aware of the powerful narrative of Western modernity that sees the liberation of the state and knowledge from the control of the church—by fencing religion out of the public domain—as the defining forward momentum of the modern secular age. Whilst the liberations of modernity are genuinely more complex than a liberation from religious wars and the trespassing of the church into secular affairs,[18] the Pope deeply appreciates that the Catholic Church has many past and present political sins it needs to repent of. And yet, *Laudato Si'* contains a subtle critique of the modern idea of religion itself, a carefully calibrated rejection of the separation of nature from grace, and it seeks to insert Christian theology back into the realm of public action, public morality, and what is largely seen as the entirely secular and practical realms of commerce, employment, technology, and politics.

Two things are apparent here. Firstly, secular power remains deeply formed by PDT. Secondly, a significant feature of that deep formation is the isolation of the sacred from the secular. Here, theology itself becomes unable to speak to theological error embedded within the public domain. But the situation is even more difficult. Secularised PDT has become convinced that it is not theological and, thus, has nothing to learn from theologians. This has further implications.

There can be a "godless" secularism where a pragmatic consumer domination logic with no transcendent or essential moral horizon, tacitly rejoices in its offence to Christian theology. Ironically, such "godless" secularism has deeply theological and Christian roots.[19] Equally, there can be a "godly" secularism where theology is firmly cordoned off from the pragmatic consumer domain to protect religious freedom. Company board members of large fossil fuel corporations, of coal mining corporations, of industries like armaments, may be personally religious and moral individuals, but as company directors in the secular commercial arena, it is the profit interests of the company that are the first defining drivers of their professional choices. This is the separation of religion and personal morality from practical business, technological feasibility and legally required business ethics. Once this separation is in place, religion itself becomes entirely individually optional and discretely isolated from power and realism. Religion is, thus, the enemy of Christian theology when it comes to any public action a Christian may want to perform.

In reality, however, secular PDT is a political theology, but it is a modern secular political theology that refuses to recognise itself as a theology. This means it is profoundly protected from the only type of critique that would de-legitimate it—a genuinely theological critique. The Guardini/de Lubac critique of secular modernity, including their critique of the idea of a discrete realm of religion, lies behind Pope Francis' deft ability to not compromise his stance by accepting a harmless and merely spiritual role for religion.

To take intrinsic values and transcendent meanings seriously in the way we actually live, in how we relate to ourselves, each other, and the world, is to be human. It is a failure of humanity, a lack of proper humanism, that upholds an amoral realism in the commercial, political, and intellectual culture of our global civilisation. This is the Pope's message, and should we heed it—or even if only 1.3 billion Roman Catholics heeded it—then Catholics might recover some of the lost public ground that "religion" has taken from the church, and better serve truly human interests for the common good.

The underlying aim in *Laudato Si'* is nothing short of the conversion of the cultural norms that legitimate global corporate, technological, military, financial, media, and political power. In this, there is hope. Without such a conversion, what climate scientists call "business as usual"[20] will irreparably vandalise our common home.

In important regards, the Roman Catholic Church has never really believed in the modern secular conception of religion. This gives them some distance from the secularised theological assumptions of modernity. This comment segues neatly into a quick look at other forms of Christian theology that are not native to Western modernity, that equally have eco-theological implications that can challenge PDT assumptions and practices.

5.2 Celtic, Orthodox, and Indigenous Christianity

Thanks to the existence of the Vatican, it is considerably easier to locate what an authorised contemporary (be it continually unfolding) Roman Catholic stance on ecological issues is than it is to locate what any sort of decisively authorised Celtic, Orthodox, or Indigenous Christian stance on eco-theology is.[21] This is a problem well recognised in the literature.[22] So I will here only make quick reference to three relatively recent examples of Orthodox, Celtic, and Indigenous Christian texts relevant to eco-theology, rather than making any strong claims about what the larger categories might be reasonably described to be.

PDT is modern and Western, but all Christian traditions—except the Evangelical non-conformists—predate this period and so have solid resources within their heritage that provides them with the means of distancing themselves from the environmentally destructive features of Western modernity. The Roman Catholic tradition—as well as the Reformed traditions of the sixteenth century—were certainly shaped by the late medieval nominalist, voluntarist, rationalist, and materialist[23] innovations that were vital for the appearance of the modern sensibility in the seventeenth century. And yet, these traditions have longer and deeper roots than these fourteenth-century innovations, that can be brought to life again, and that strongly challenge PDT. Christian traditions that have never been strongly modern and Western are of interest in this short section.

5.3 Celtic Eco-Theology

As Cahill points out, Celtic monasteries in the so-called Dark Ages largely produced medieval Christendom.[24] The Celtic monks were certainly Celtic, but some were also scholars of the Greek learning of antiquity that had largely disappeared from the continental domain of the old Western Roman Empire. John Scotus Eriugena's ninth century work *On the Division of Nature* (*Periphyseon*) is an astonishing synthesis of Christian theology with Neoplatonist philosophy, informed by patristic theology.[25] The point I am making here is that as early as the ninth century, Celtic Christians had long been thoroughly Christian, thoroughly embedded in Classical and Patristic thinking, and if they retained any pre-Christian magical/pagan conceptions of nature, that was fully immersed in, and reinterpreted by, the Christian scriptures, Greek philosophy, and patristic theology.

Significantly—and no doubt to the Continental West, troublingly—the theology of creation in great Celtic thinkers such as Eriugena is theophanic; though one does not worship creation, creation is a sensible manifestation, a veiled appearance, a participatory icon of God's word and presence. In Eriugena's *Periphyseon*, uncreated reality—God—is distinct from the created nature of transcendent forms and of angelic, temporal, and physical beings, even though all created being comes from God, depends entirely on God, and returns to God. Created being's highest purpose is to reveal the glory of God. This means one cannot have a merely instrumental approach to nature, and this means that what we might call natural magic—the divinely touched nature of creation, and even the strangely animate nature of inorganic reality and the spiritual quality of the natural environment itself—retains a stronger hold on the Celtic Christian mind than it does on the Latin Christian mind. Certainly, some contemporary Celtic Christians, such as Mary Low, uphold a theological vision of the earth that is in some manner a theophany of the living God.[26] A reverential and non-instrumental respect for nature, then, is seen by Low as an integral part of Celtic Christianity. Unsurprisingly, the suspicion that any Christian form of re-enchantment of the natural world is both an invitation to the idolatry of nature worship, as well as an attack on modern science, is felt very keenly by classically modern Christians.[27] We shall look at this more closely when we look at Evangelical eco-theology. But clearly, Celtic Christian eco-theology (whether one considers it genuinely Christian or not) certainly does provide an alternative to PDT.

5.4 Orthodox Sacramental Theology and Western Attitudes to Science

Since becoming the 270th Archbishop of Constantinople in 1991, the Ecumenical Patriarch Bartholomew has been known as the Green Patriarch. His statement from Santa Barbara in 1997 shows the urgency and seriousness of Orthodox eco-theological concern:

> To commit a crime against the natural world is a sin. For human beings to cause species to become extinct and to destroy the biological diversity of God's creation; for human beings to degrade the integrity of the earth by causing changes in its climate, by stripping the earth of its natural forests, or by destroying its wetlands; for human beings to injure other human beings with disease by contaminating the earth's waters, its land, its air, and its life, with poisonous substances – all of these are sins.[28]

The relationship between ecological crisis and theology strongly exercises contemporary Orthodox theologians, such as Metropolitan John Zizioulas.[29] Understandably, the way Orthodox theology approaches nature is very different from a Western Christian approach. It is the centrality of a sacramental understanding of reality as embedded in the expansive meaning of the Christian Eucharist which gives Orthodoxy its distinctive eco-theology. This theological way of seeing creation is richly outlined by Alexander Schmemann's classic text *For the Life of the World*.[30]

The leaders of Orthodox Christianity today are in no sense climate science sceptics. Even so, their attitude to science itself is quite different to a Western more instrumental and reductive outlook. For whilst the Orthodox have no inherent fear or antipathy towards modern science, and are as pleased as the next person about the conveniences of jet travel and the life-saving powers of penicillin, the Orthodox perspective Schmemann outlines just doesn't think science has anything to say about that which is really primary; the sacraments and the Life of Christ given to the world through the incarnation and passion of Christ, as expressed right now in the Eucharist. The Orthodox are not Western in that they do not see the world through a nominalist, a voluntarist, and a pure matter theological lens. Consequently, the Orthodox do not have such an instrumental and "activist" approach to nature as Western Christendom does. Orthodox Christianity has been criticised as being too poetic, too mystical, and even too magical to produce modern science, but perhaps there is some truth to this, and it is a strength rather than a weakness. That is, the sensitivity to theologically framed ontology as grounding the lived experience of the life of the Church, the sense that instrumental power and epistemic mastery are all rather secondary to that which is essential to the real living of Christian faith, this is something that Western Christianity is weak in, but that weakness can be addressed by Eastern Christianity.

And indeed, patristic (that is, orthodox) roots run deep in Western Christianity, particularly in the theology of sacraments and in a sacramental understanding of creation. This is not only true in Roman Catholic theology, but also in non-Zwingli inflected forms of Protestant theology, such as trajectories within Lutheran, Anglican, and Wesleyan theology. In this regard it is, I think, important to nuance Lynn White's thesis, in that "Western Christian theology" is a complex and multi-trajectory animal that retains Eastern Christian trajectories within it, particularly in its sacramentally inflected understandings of creation. Indeed, the impact of Wesleyan revivalism on

Evangelicalism remains an open door to sacramental theology even within very modern and Protestant forms of Western Christianity.[31] The work of Evangelical scholars such as Robert Webber[32] and Hans Boersma[33] in explaining patristic theology to Evangelicals is motivated by Webber's conviction that modern Western Christians will become more faithful to Christ if they become less committed to Western modernity.[34]

5.5 Indigenous Christian Eco-Theology and Non-Western Epistemologies and Ontologies

As Philip Jenkins points out, the way the peoples of the Global South read the Christian bible is very different from how modern Western Christians—be they Liberal or Conservative—read the bible.[35] Non-Western people typically do not see a sharp separation between the natural and the super-natural assumed in Western modernity, they do not see the autonomy of the individual from the collective the same way that modern liberal consumers do, and they do not think of "religion" as a defined domain limited to institutionalised worship, personal convictions, personal morality, and unit family morality. That is, they read the bible from worldview categories far closer to those that the bible itself is written in than do modern Western people.

Then there is power and status. People dispossessed, culturally devastated, structurally impoverished, and racially marginalised within Western modernity by the aftermath of the high age of European colonialism, have no innate commitment to the impossibility of today's normal power relations fundamentally changing. Indeed, using Brueggemann's categories, it is precisely to such people that the prophetic imagination of the Hebrew and Christian scriptures speaks most directly. Many in the Global South are not embedded in the epistemic and instrumental norms that govern the dominant order and so the radical message of the outcast and crucified Christ is an "energizing hope [that] comes precisely to those ill-schooled in [the dominant categories of] explanation and understanding".[36] If anyone is going to think outside of the box of the necessary power and privilege categories that define the present global order, it is thinkers from the Global South and Indigenous thinkers. Indeed, such thinkers have good reason to be highly suspicious of the categories of scientific objectivity, progress, and modern instrumental power.

The enormous craze in phrenology[37] in the mid-nineteenth century entailed measuring Indigenous heads and stealing Indigenous skulls in Australia, supposedly scientifically demonstrating why these "primitive" races were simply not adapting to European-imposed civilisation.[38] Early in the twentieth century, the Commonwealth and States of Australia typically assumed a Malthusian and Social Darwinist understanding of the massive population decline of Australian Aboriginals after white settlement. It was "assumed that the disappearance of the native race [was] some kind of final solution".[39] As a result, "it became usual for a protector to remove children with light skins from their mothers."[40] "Light-skinned" Aboriginal children might survive and be integrated into white Australia, but full-bloods were destined to die out, by natural selection. Removing Aboriginal children from their families by the state was, thus, normal legal "protectorate" policy in Australia from 1905 to 1967. When you have been the *object* of modern Western knowledge and instrumental state power, one tends to have strong tacit sympathy with a somewhat Foucauldian understanding of knowledge.[41]

The Australian Aboriginal thinker Tyson Yunkaporta recently published a provocative book with the subtitle "How Indigenous Thinking Can Save the World".[42] It's a fascinating book. The impact of Western thought, religion, technology, and power is deep on Australian Aboriginals, and a great deal of it has been shattering, not just of the old ways of the first nations, but of modern non-indigenous Australians (and Australia) as well. Thinkers like Yunkaporta can navigate two worlds and reflect indigenous understandings of knowledge, being, humanity, and nature back to the West in ways that are astonishing, window opening, and may in fact be just what we need to "save the world." Indigenous thinking's contrast with the Western determination to harness and adapt nature, rather than adapting ourselves to nature, is one area that may well speak to our climate change crisis very directly.

Yunkaporta also points out that it is not academic systems that one can study abstractly that he has to offer; his indigenous *experience* of relationships, of the world, of different orders of spirit and nature interaction, informs the indigenous ontological wisdom and the indigenous epistemological framework that he is a custodian of. Whilst Christian missions must have touched his kin in many places,[43] Yunkaporta is not a Christian theologian. Without having to navigate a modern and white fear of Indigenous knowledge and spiritual experience in theological seminaries, Yunkaporta can think and feel as an Aboriginal in ways that are only just beginning to gain safety in most Christian organisations and seminaries in Australia. Yet, Indigenous Christian theologians in Australia equally want to explore what it is that is distinctive and different in their theology to modern and Western Christian theology.

Ken Lechleitner Pangarta—a Central Australian Aboriginal man from West Aranda, and a Lutheran—asks the question, "Was God present in Australia before 1788?"[44] Pangarta goes on, "If the bible is to be believed (Ecclesiastes 3:11, Acts 17:22–29, even Hebrews 1:1–2), then the answer is a resounding 'Yes!'"[45] From this biblical opening Pangarta goes on to explain that his ancestors worshipped God in a way that was fulfilled—as Christian Aboriginals see it—in the Christian gospel. Theological continuities between the old ways and the Christian gospel are plain to Pangarta, such as seen in the meaning of the traditional Aboriginal word for God, Altjirra. Pangarta explains.

> Both God and Altjirra are self-revealing ... Both come with their own narratives, both are seen as creative of all that exists. Both have revealed, through their various covenants, the ways and means for people to exist and lead meaningful and fulfilling lives.[46]

The point to notice here is that Indigenous "religion" (not an adequate word at all) is certainly no less theologically rich than Western religion, and perhaps more ontologically akin to Eastern orthodox conceptions of being and Celtic Christian conceptions of nature than is modern Western theology. There is nothing "pagan" about Pangarta's Lutheran Indigenous theology of Altjirra. And when it comes to eco-theology, the notion of being caretakers of land—of managing it, and nurturing it with care, for we depend on it for our flourishing—is deeply embedded in the traditional understanding of divinely given Law. Pangarta points out that like the Old Testament, duty to God is deeply tied up with land-care:

> We depend on the environment for life. Our land is the gift of God/Altjirra to us. Our Law tells us what to do with it, just as Biblical Law does.[47]

As John Harris points out, apart from some outstanding missionaries, English-speaking Australia's woeful linguistic understanding of the stunningly complex Indigenous languages[48] has been a deep hindrance in understanding traditional Indigenous theology, wisdom, ethics, and philosophy.[49] The complete mistranslation of *altjirringa*—meaning something like "from eternity" or "connected with eternal things"—as "dreamtime" is a tragic case in point.

Christian Australian Indigenous theology is deeply embedded in kin-bonded, pre-monetary, richly adaptive to country, and entirely pre-secular practices of life. This is a profoundly spiritually engaged form of life with a rich ontological theology and theologically sophisticated epistemology integral with the very fibres of Indigenous experience and practice. This theological way of understanding nature could not be more different from PDT.

5.6 Mainline Eco-Theology

So far, this chapter has been discussing theologies of nature that reject PDT by looking to pre-modern or non-Western Christian theologies of nature. Chapter 6 will look at one family of Christian theology that largely upholds various forms of PDT. In terms of that axis of contrast, Mainline eco-theology functions on a rather different axis. Mainline eco-theology tends to largely reject PDT, as well as often maintaining a strongly modernist outlook.

Tensions within Mainline eco-theology seem to run along an axis of debate defined by *orthodox* Lutheran, Calvinist, Wesleyan, Anglican, Presbyterian, etc. theologies of nature, as contrasted with *Liberal* Lutheran, Calvinist, Wesleyan, Anglican, Presbyterian, etc. theologies of nature. Liberal theologies of nature can look remarkably similar to Progressive and non-Christian theologies of nature, whereas orthodox Protestant theologies of nature can be classically modern — in the pre-nineteenth century sense — or they can appeal to the more patristic and catholic heritage of the Reformers and Wesley.

I must apologise for largely by-passing Mainline eco-theology in this text as it is the axis defined by the opposition of the Christian pre-modern/non-Western theological pole versus the Christian pro-modern Western theological pole that I am seeking to unpack. Yet, though it is entirely inappropriate to equate Mainline with Liberal, it is true that many Mainline theologies of nature are theologically Liberal, and—in terms of the axis this text explores—something should be said about Liberal Christian theologies of nature.

Liberal—as a theological term—largely means strongly influenced by nineteenth-century German Protestant thinking. Here, modern science was applied to the understanding of scripture, history, and theology, with characteristic Germanic intellectual rigour and completeness. Kant and Hegel have a powerful influence over the ethics and politics of German Liberal thinking as well.

German Protestant Liberal theology has a high academic pedigree. Going back to thinkers like Friedrich Schleiermacher and David Strauss, Liberal Christian theologians have always had a deep sensitivity to myth. They have also synthesised the modern methodological atheism of an objective rationalist and empiricist outlook with what they see as the higher meaning of the Christian religion, which is its mythic and perhaps (at least for some) mystical meaning. They are a highly modern form of Western Christianity, and as such, they are open to the Epicurean, animist, eco-feminist, and eastern theologies of nature that are re-appearing in Western intellectual circles. But this is a complex terrain. Whether "panentheism" is really animist, and whether a carefully nuanced materialism is really incompatible with Christian doctrine, are not matters that are easily determined without demanding and careful intellectual exploration.

Mainline eco-theology, then, can be Liberal, modern, or orthodox, or any combination thereof. It is often on the Progressive spectrum politically. Most Mainline eco-theology that I am familiar with rejects PDT and is seeking to promote climate change mitigation, and politically framed ecological reform.

Notes

1 Interestingly, there are intellectually powerful Christian theologians across a range of ecclesial traditions who cogently argue that the usual way we think of "theology," "religion," and "Christianity" are now enormous problems. See Liethart, "Against Theology" in *Against Christianity*, 45–75; Ellul, *The Subversion of Christianity*; Yannaras, *Against Religion*; Budde, *The (Magic) Kingdom of God*; Hauerwas, *After Christendom*.
2 Francis, *Laudato Si'*, 7 [1]. (Note: the numbering here is that of the English booklet version, followed, in square brackets, by the paragraph number.) It is interesting that Lynn White, in 1967, appealed to St Francis as a figure who might inspire a radical change of mind for our civilization.
3 Francis, *Laudato Si'*, 9 [6]. "The deterioration of nature is closely connected to the culture ... [where] human freedom is limitless."
4 Francis, *Laudato Si'*, 9 [6]. "... creation is harmed ... where everything is simply our property and we use it for ourselves alone."
5 Francis, *Laudato Si'*, 9 [6]. "Man does not create himself. He is spirit and will, but also nature."
6 Francis, *Laudato Si'*, 9 [6]. "... the misuse of creation begins when we no longer recognize any higher instance than ourselves ..."
7 Francis, *Laudato Si'*, 15 [19].
8 Francis, *Laudato Si'*, 17 [23].
9 Francis, *Laudato Si'*, 20 [29].
10 Francis, *Laudato Si'*, 21 [32,33].
11 Francis, *Laudato Si'*, 29 [53].
12 Francis, *Laudato Si'*, 35–36 [67–68].
13 Francis, *Laudato Si'*, 35 [66].
14 Francis, *Laudato Si'*, 39 [75].
15 See Guardini, *The End of the Modern World*. See Francis, *Laudato Si'*, 53–59 for five direct citations of this text by Guardini in the Pope's notes.

16 Rowland, *Ratzinger's Faith*, 19–29.
17 De Lubac, *The Drama of Atheist Humanism*.
18 See Cavanaugh, *The Myth of Religious Violence*.
19 See Milbank, *Theology and Social Theory*; Gregory, *The Unintended Reformation*.
20 Maslin, *Climate Change*, 134.
21 Of course, Roman Catholic theology is anything but monolithic. Even so, it does have a centralized teaching authority, and the encyclicals of the Pope provide a broad but authorized account of what the current state of doctrinal understanding is.
22 Take "Celtic" Christianity for example. See Meek, *The Quest for Celtic Christianity* which in part is a response to the fascinating book by Low, *Celtic Christianity and Nature*.
23 By materialist here I mean the development of complex thought regarding the ontological self-standing of ungraced matter. This medieval pure matter trajectory is vital for the appearance of reductive materialism in Western thinking, which only really comes into its own in the nineteenth century. For the medieval roots of our materialist tradition, which definitely are Christian, see Pasnau, *Metaphysical Themes*, 47–52.
24 Cahill, *How the Irish Saved Civilization*, 145–196.
25 John the Scot, *Periphyseon* [This was Latinised in the seventeenth century as *De Divisione Naturae*, i.e., *On the Division of Nature*]. In this brilliant and demanding dialogue, Eriugena's explores the four divisions of *physis* (i.e. nature) which are these: 1. that which creates and is not created (God); 2. that which is created and creates (Platonic forms); 3. that which is created and does not create (the created world, including us); 4. that which is neither created nor creates (God again, the alpha and omega of all that is). The Plotinan *exitus et reditus* (exit and return) of reality from the divine One (God) and back again is the philosophical framework of Eriugena's speculations here. There has been considerable debate over the centuries whether this makes Eriugena a pantheist (which he denies, but Pope Honorius III in 1225 considered him so) or, in more contemporary categories, panentheist, but Western appraisals of his thought seem prone to reading him in ways that are not sensitive to his more Greek influenced theological perspective. Even so, the orthodoxy of his commentary on the Gospel of John is generally unquestioned, though it contains strong Neoplatonist themes, and there was no thinker in the West in the ninth century equal to his learning and intellectual power.
26 See Low, *Cherish the Earth*. This is an anthology of eco-theological songs and poems from many traditions and religions, but with a strong Scottish and Irish emphasis.
27 See Tyson, *Seven Brief Lessons on Magic*. In this text I have explored different philosophies of nature, and I have described those things that science cannot account for – thought, love, purpose, beauty, meaning, etc. – as "magic." This has not been terribly well received by modern Christians in the scientific fraternity.
28 As cited by Chryssavgis, in "Ecumenical Patriarch Bartholomew: Insights into an Orthodox Christian Worldview," 9–18 https://www.patriarchate.org/the-green-patriarch.

Non-Modern Christian Theologies of Nature   93

29 My thanks to Sotiris Mitralexis for drawing my attention to these lectures by Zizioulas: "Preserving God's Creation: Three Lectures on Theology and Ecology," 1–5, https://www.resourcesforchristiantheology.org/preserving-gods-creation-1/; "Preserving God's Creation: Three Lectures on Theology and Ecology," 41–45. https://www.resourcesforchristiantheology.org/preserving-gods-creation-2/; "Preserving God's Creation," 1–5. https://www.resourcesforchristiantheology.org/preserving-gods-creation-3/.
30 Schmemann, *For the Life of the World.*
31 See these intriguing texts: Cross and Thompson, *Baptist Sacramentalism*; Harmon, *Towards Baptist Catholicity*; Harvey, *Can These Bones Live? A Catholic Baptist Engagement with Ecclesiology, Hermeneutics, and Social Theory.*
32 Webber, *Ancient-Future Faith.*
33 Boersma, *Seeing God.*
34 Weber, *Ancient-Future Faith,* 18–33.
35 Jenkins, *The New Face of Christianity. Believing the Bible in the Global South.*
36 Brueggemann, *The Prophetic Imagination,* 104.
37 Phrenology was the study of skull shapes, supposedly differentiating "high" from "low" races.
38 Thearle, "The Rise and Fall of Phrenology in Australia."
39 Rowley, *The Destruction of Aboriginal Society,* 223.
40 Rowley, *The Destruction of Aboriginal Society,* 231.
41 Foucault, *Discipline and Punish. The Birth of the Prison.*
42 Yunkaporta, *Sand Talk.*
43 See Harris, *One Blood.* A common pattern of white settlement in Australia was Aboriginal dispossession, murder, rape, disease, and then missions. The gospel of Christian love was, unsurprisingly, hardly rapidly embraced under these conditions. It is clear that Aboriginals were by no means mere victims of White settlement either; resistance and adaptation are evident on every front. See Reynolds, *The Other Side of Frontier.* The story of Christian missions Harris outlines is deeply complex. Christian missions were part and parcel of colonial Australia, and the agony of that legacy is deeply acknowledged by Harris. But it is also the case that were it not for Christian missions, genocide may have been achieved. As a result, Aboriginal Christianity is a strong feature of contemporary Indigenous Australia, touching most Indigenous families in Australia.
44 Pangarta, "Central Australian Aboriginal People's Worldview on God/Altijirra," 3. 1788 is the year the first English settlement was set up in Australia.
45 Pangarta, "Central Australian Aboriginal People's Worldview on God/Altijirra," 3.
46 Pangarta, "Central Australian Aboriginal People's Worldview on God/Altijirra," 3.
47 Pangarta, "Central Australian Aboriginal People's Worldview on God/Altijirra," 4.
48 German translators did a much better job in doing the serious linguistic work to better understand traditional Indigenous theology. The Lutheran missionaries Herman Kempe and Wilhelm Schwarz took the effort in the

late 1870s to learn Aranda, Ken Pangarta's mother tongue. Without hesitation they used the divine name Altjirra, who was eternal and entirely good, as a theologically appropriate name for God. See Harris, "Seeing God," 6.
49 Harris, "Seeing God," 5–9.

6 Contemporary Christian Theologies of Nature and Climate Change
Evangelicals

Unlike Roman Catholic, Eastern Orthodox, Celtic, and Indigenous forms of Christian eco-theology, we now come to an archetypally modern and Western way of being Christian. Evangelicals are born in the modern age, they are great promoters of modern secularism and modern religion, and the theological commitments that undergird their own approach to technology and to liberal, democratic, free-market consumerism, are entirely modern. There is an intimacy in the relation of Evangelicals to modern PDT, but even so, this is a very complex theological animal.

Evangelicals are committed to various interpretations of the "plain meaning" of the authority of Scripture and read its meaning through modern eyes where a more or less scientific positivism is the assumed truth criteria. That is, if the Bible reports a miracle such as the resurrection of Christ, this would be a verifiable physical resurrection from literal death, not a metaphor, not a myth, not some mistaken misunderstanding of what really happened. Evangelicals reject any non-miraculous understanding of the resurrection. With Saint Paul they would maintain that if Christ did not really die on the cross, and if Christ did not really physically rise from the dead, then the Christian faith is complete nonsense.

Evangelicals usually believe in an entirely natural and non-sacred nature, in a discrete supernatural reality, and in supernatural miracles that interrupt natural reality. This is consistent with the theological outlook of many of the English-speaking early modern pioneers of science.

Due to an affinity with small business entrepreneurialism and sympathy for libertarian and eighteenth-century classical economic conceptions of free-market capitalism, the majority of white American Evangelicals[1] tend to have right leaning political commitments.[2] The revivalism native to Evangelicals is often also very much at home with emotively persuasive forms of public communication, which readily translates into an ease with advertising, entertainment, and consumer culture. Evangelicals fully belonging to liberal democratic and secular modernity, as seen in the United States, and one could claim that the "American Way" is deeply indebted to them, particularly in the realm of religious freedom and individualistic liberalism in general.

One thing that is a bit difficult to appreciate about Evangelicals—particularly since the Scopes Trial[3] and the birth of anti-Darwinian fundamentalism in the early twentieth century—is how science and technology friendly Evangelicals are. Many twentieth-century American Evangelicals became radically opposed to Darwinian evolution,[4] but they did so not out of some reactionary disdain for modern science but out of what they saw as their own commitment to a scientific reading of nature through divinely authorised scripture.

Evangelical fundamentalism was born in reaction against the Liberal Protestant mythic interpretation of core Christian doctrines. Prior to the late nineteenth century, an integrative though methodologically distinct approach to treating the bible as the highest and divinely revealed source of historical truth, and treating science also as a sure source of natural truth, was normative in Western culture. But this norm was radically reworked as reading the bible through the lens of scientific naturalism became increasingly embraced by Liberal Protestant theologians in the nineteenth century. Progressive materialists denied the reality of the supernatural entirely and this filtered into mainstream Western intellectual culture by the turn of the twentieth century. In contrast, Evangelicals saw the denial of the reality of the supernatural, and the application of scientific naturalism to scriptural hermeneutics in a manner that mythologised traditional creedal beliefs, as heresy.

Twentieth-century Evangelicals were Conservative, certainly, as regards nineteenth-century science-and-bible norms. But this Evangelical integration of the bible and science has no relation to Western pre-modern conceptions of reading Nature and Revelation together, for medieval thinking was much more allegorically and metaphysically complex than modern hermeneutics, and was tied to Aristotelian natural philosophy. "Creation Science" is a distinctly modern approach to integrating the book of Nature with the Christian Scriptures. Evangelicals—be they six-day creationists or old earth theistic evolutionists—are usually firm modernists as regards knowledge and nature itself.

We will now look quickly at three texts by Evangelicals on nature and on different aspects of what we now call the green movement.

Modern Christian Theologies of Nature 97

With the aid of Veldman's careful scholarship on Evangelicals and climate change,[5] these texts will show that Progressive Dominion Theology is a deep but complex driver of Evangelical thinking as regards environmental concerns.

6.1 Francis Schaeffer's Evangelical Creation Care Theology

Francis Schaeffer was an ordained Presbyterian minister from Pennsylvania. In 1955, he set up a drop-in community in Switzerland called L'Abri ("the shelter") that was designed as a place of refuge for searching counter-culture youth. L'Abri also aimed to be intellectually and artistically stimulating for young Evangelicals trying to make sense of the post-war world without abandoning a traditional Evangelical understanding of Christian scriptures and theology. Schaeffer was also intimately involved in the politicisation of Conservative Evangelicals as he persuaded Jerry Falwell to start up the Moral Majority movement in 1980. The Moral Majority attracted key Conservative Evangelical leaders and formed ties with the Republican Party to advance a Protestant and Catholic campaign against legalised abortion.

In 1970, Schaeffer published *Pollution and the Death of Man*.[6] Prominent Evangelical thinkers like Schaeffer in the 1970s showed deep empathy for the questions and concerns that characterised the youth movements of the 1960s.[7] Schaeffer had profound sympathy for the existential anxiety posed by the Cold War threat of nuclear annihilation; he empathised with environmental concerns about natural resource exploitation that had no regard for preserving natural beauty; he was concerned about the destruction of rare and delicately balanced ecosystems; he was grieved by the extinction of species. That is, many substantive concerns of what we could call the Progressive Green Left were shared by Schaeffer. It now seems somewhat dissonant that this counter-culture friendly, romantic nature lover, and environmentally sensitive figure became a key player in the alignment of Conservative Evangelicals with Republican Party politics.[8] The historical logic of how this happened is illuminating.

In *Pollution and the Death of Man*, Schaeffer accepts the moral and aesthetic urgency of addressing the environmental issues of his times,[9] yet he was strongly critical of Lynn White's assessment that Western Christian theology causes our ecological crisis. Schaeffer admired White for bringing first-order concerns to the fore when thinking about the environment, but setting himself against White, Schaeffer maintained that it is the *rejection* of a genuinely Christian theological understanding of humanity's stewardship of creations that causes our ecological crisis.

As Schaeffer saw it, the central threat to the cultural and ideological integrity of the United States as a Christian nation—and hence, to its capacity to uphold human flourishing—was secular humanism. Rejecting God and placing Man at the centre of knowledge, power, and purpose leads, in Schaeffer's view, to the death of both Man and Nature, rather than to liberation and progress. True freedom and true progress happen only "under God", where proper deference to the Creator enables proper deference to creation and one another. In Schaeffer's view, the cultural revolutions of the 1960s are the result of a secularised eschatology of progress based on scientific naturalism and an idolatry of the worship of Man.

Schaeffer held that a post-Christian religious consciousness was replacing a genuinely Christian understanding of both Man and Nature in twentieth-century America. As Schaeffer reads it, the environmental movement correctly recognised that things were going wrong, but—thanks to White—incorrectly attached the cause of that wrong to exactly the wrong quarter: Christian theology. Schaeffer saw the emerging post-war Green movement as a spiritually hungry movement, often dissatisfied with an instrumental scientific naturalism, so they sought to replace a Christian theology of creation with a turn to Eastern pantheism and indigenous animism. In contrast to this turn, Schaeffer seems to support Butterfield's understanding of the positive meaning of an anthropology where Man is, in significant regards, apart from nature. To Butterfield, a Christian understanding of history holds the human world as exactly *not* defined by the cruel brutality and blind indifference of nature. Animism and pantheism—as well as modern scientific naturalism (taken as a meaning framework)—seem inherently prone to a tendency towards brutality and indifference. Thus, when human meaning has no genuinely transcendent frame of reference beyond nature, a naturalistic brutality becomes the realist bedrock of power, and this produces Nazism and Marxism that have no regard for the intrinsic dignity of individual persons, no regard for any frame of moral reality above military and economic power, and no other logic than callously indifferent instrumental reason.[10] For these reasons—which are outside of the scope of the argument that White actually puts forward— Schaeffer holds Lynn White to be profoundly theologically wrong, and Schaeffer rather blames White for Christianity being falsely labelled as the causal villain of the ecological crisis of our times.

Schaeffer's reading of White is at considerable cross purposes to White's actual argument, so it is not a particularly fair reading of White. And yet Schaeffer's argument itself—given his theological commitments—is by no means facile. If one believes in a transcendent personal God, and that the traditional "American Way" is embedded in shared beliefs and practices premised on the reality of a transcendent personal God, then Schaeffer's diagnosis of post-war America is hard to ignore. At least, it is hard for about 100 million Evangelicals in the United States to ignore.

From where Schaeffer sits, he sees all manner of things going wrong in his times: a new paganism is rising through the environmental movement; the sexual revolution of the 1960s has promoted promiscuity and deeply undermined Christian marriage and the cultural security of an assumed code of inter-gender chivalry; gay rights, militant feminism, abortion, and the removal of prayer and bible reading from state schools in the United States are all symptoms of a cultural rejection of biblical revelation and turning to an anti-Christian secular humanism; art and music have become overly sensualised and their transcendent horizon has been degradingly immanentised; and there is a turn by the youth to eastern mysticism, eastern philosophy, and drugs in their search for meaning as they grapple with the spiritual vacuity of hedonistic consumerism.[11] This larger picture of cultural and theological concern is the context in which Schaeffer's understanding of creation care is situated.

To address the above concerns, Evangelicals got political. The Moral Majority movement of the 1980s got busy and forged the political, promotional, lobbying, and financial ties that matured into a firm alliance between the Christian Right and the Republican Party. It is important to realise that key founding figures in this movement—such as Schaeffer—upheld a firm commitment to ecological concern. But as things developed, the dynamics of ecological politics became highly complex. Powerful trajectories of influence were moving in opposite directions. By the turn of the twenty-first century, highly influential Evangelical voices were promoting the climate change denial trajectory, such as James Dobson and Jerry Falwell. In contrast, on the climate change mitigation trajectory, Al Gore (Democrat Vice President under Bill Clinton, and a Baptist) was working hard to promote climate activism to all. In the first decade of the twenty-first century, Al Gore—with left-leaning Evangelical conservationists, with new generation Evangelical eco-theologians like Jonathan Merritt, and with science-converted Conservatives like Richard Cizik (converted by John Houghton)—was working hard to promote an "Evangelical greening".

In this strongly contested environment, the decisive victory of denial over mitigation only really emerges during the 2008 Global Financial Crisis. There may well be an intimate relationship between the 2008 shaking of the post-Reagan era financial sector in a context where terrorist fear was already a strong political motivator, such that a heightened leverage for fear based denial was perhaps irresistible to the polity of the United States. But for whatever reason, the "Evangelical greening" was politically defeated just as the Democrat, President Obama, came into office. As one example of how the Evangelical greening was defeated, Veldman describes the institutional blowback in the Southern Baptist Convention against the climate change action plan put forward by Jonathan Merritt.[12] In this instance, the combination of the Evangelical Climate Initiative put forward in 2007 and Jonathan Merritt's 2008 declaration, galvanised the powerful and lobby-savvy institutional machinery of the Southern Baptist Convention against climate change mitigation.

Yet, climate change denial is *not*, in itself, a doctrinal cornerstone of the Religious Right. To the contrary, theological concern for the environment *is* a doctrinal cornerstone of Christian creation care theology, but it is coupled—to Evangelicals—with a deep suspicion of the theological and moral commitments of the secular environmental movement. This is all there to see in Schaeffer, and this makes the tripability of Evangelical environmental concern so delicate. Being as inherently delicate as the balance towards or away from climate change action is—and, as theologically nuanced as it actually is—it could tripped back to an activist climate change mitigation stance. As this would have profound political implications, it bears closer examination.

6.1.1 Schaeffer and the Philosophical Delicacy of the Evangelical Stance on Environmental Action

Reading Schaeffer's *Pollution and the Death of Man* it is clear that the sort of concerns he has are philosophical theology concerns in relation to their social and political implications, and that the interlocutors he opposes (such as Lynn White) are more springboards for his own ideas than serious intellectual partners.[13] Putting White to one side for the moment, what, then, is Schaeffer himself saying in *Pollution and the Death of Man*?

The first thing to notice is that Schaeffer, like theologically Conservative Evangelicals in general, has not made the hermeneutic transition of the late nineteenth century when science became the first truth lens of

Western modernity through which religion and everything else was then viewed and interpreted. In this regard Evangelicals are conservative modernists who view science and everything else though the hermeneutic lens of their reading of Christian scripture/theology. Contemporary Evangelicals often see themselves in the same tradition as Christian scientists like Robert Boyle and Michael Faraday. And whilst Evangelicals definitely are *modern* Christians (the first modern Baptist dates back only to 1609) in terms of this hermeneutic commitment, they are simply orthodox. The orthodox Christian outlook has always maintained that the Christian revelation and the practice of Christian life is the first truth discourse for the Christian, and natural philosophy (science) is of secondary significance (though not unimportant) in relation to those truths that are of first importance. The most pernicious enemy of a Christian understanding of truth to a traditional Evangelical, then, is Liberal Protestantism, wherein science has become the truth lens through which religion and faith are interpreted. Evangelicals such as Schaeffer tend to assume that a Liberal truth lens in theology and a secular humanist lens in philosophy impose at least a functional reductive naturalism, which—again, at least functionally—excludes orthodox Christian doctrine from the realm of public truth. There is nothing misguided about Evangelicals thinking this is the case. Equally, the dogmatic insistence that reductive naturalism indeed does require Christians to keep their theological faith convictions firmly quarantined from the realm of public truth is often assumed by Progressive opponents of Evangelical Christian belief and practice. Distinctive modern understandings of science, religion, and politics are thus central underlying issues that enlivens the US culture wars between Progressives and Conservatives.

Putting the authority of Scripture above the authority of secular scientific theories, American Conservative Evangelicals are not going to turn into functionally reductive naturalists any time soon. The division between Conservative Evangelicals and Liberal Protestants is a division on this very point. Liberal Protestants (and Liberal Roman Catholics) have largely embraced the early nineteenth-century hermeneutic reversal of the science and religion relationship as pioneered by German biblical scholarship, which was developed under the influence of eighteenth-century rational deism, empirical skepticism, Kantian ethics, and Hegelian historical idealism. The Evangelicals stand apart from that development and thus stand apart from the trajectory that produced modern atheism in the 1840s and the remarkable science and religion reversal in Western intellectual culture in the late nineteenth century. Evangelicals are not fools for refusing to concede to what is, in the end, an open first philosophy question about the final nature of reality.

The way a traditional Evangelical thinks about science and religion is going to be incommensurate with a studies of religion outlook that will only look at the human activities and states of mind that are—to at least some extent—transparent to the categories of modern objective knowledge. In contrast, the Evangelical sees God as the most basic grounds of the real. If many white American Evangelicals are, to put it mildly, nonconformists about evolution—though there certainly are prominent theistic evolutionists who are theologically Conservative[14]—this is because science itself is not primary to their sense of final meaning. Evangelical theological commitments mean science will be treated in a subsidiary manner to their understanding of the "plain meaning" of biblical revelation where this is deemed theologically necessary.

The reason why climate science can be approached sceptically by Evangelicals since 2008, though, is far more tied up with culture and power than with science. But significantly, there is no theological reason why an Evangelical should not be committed to taking firm action to slow down and stop greenhouse emissions. If other culturally and politically framed issues were not tweaking Evangelicals away from climate change action, there is every reason to suspect that they would be—as the Southern Baptist Jonathan Merritt put it—"Green like God".[15] We will look more closely at those cultural and political dynamics now before returning to two other Evangelical writers on the science and theology of nature.

6.2 US Evangelicals, Climate Science, and the White House

Veldman points out that in the first decade of the twenty-first century — during the Republican presidency of George W. Bush, "American Evangelicalism appeared to be in the midst of a greening trend".[16] The Evangelical Environment Network (EEN) had been established in 1993, and in 2004 the EEN's endeavours bore fruit with the National Association of Evangelicals (NEA) putting forward "creation care" as a crucial public concern. An awareness of the scientific evidence for climate change was growing in Evangelical circles. In 2006 the Evangelical Climate Initiative (ECI) published a powerful statement called "Climate Change: An Evangelical Call to Action" singed by 86 influential Evangelical leaders and published in the *New York Times* and *Christianity Today*. The four carefully defended claims in that ECI statement are these:

Claim 1: Human-induced Climate Change is real.
Claim 2: The consequences of Climate Change will be significant, and will hit the poor hardest.
Claim 3: Christian moral convictions demand our response to the Climate Change problem.
Claim 4: The need to act now is urgent. Governments, businesses, churches, and individuals all have a role to play in addressing Climate Change – starting now.[17]

In conjunction with the Intergovernmental Panel on Climate Change (IPCC), Al Gore—former Democrat Vice President and a Baptist—received the 2007 Nobel Prize for their work in public communication urging the immediate need to respond to global warming. Inertia for a US lead push for global reform looked possible. And yet, by 2008, the greening trend in US Evangelicals was coming to an end. This trend was deliberately squashed.

Climate science is not new. In 1965 the report "Restoring the Quality of Our Environment" by US President Lyndon B. Johnson's Science Advisory Committee warned that fossil fuel emission could cause serious global environmental problems if not addressed. From 1965 it was clear that solidly demonstrated science had established that anthropogenic global warming posed a serious policy problem for the White House. And yet, the specifics of what was going to happen were not certain, and the nature of the problem was too long term in scope to gain any strong leverage on policy. As one commentator has observed, in the period from 1965 to 1979, "what matters in science is not the same as what matters in politics".[18]

In 1977, "the Jasons"—a committee of elite scientists (mainly physicists) that had been advising the US government on national security issues since the early 1960s[19]—were approached by the US Department of Energy to examine their own research on the effect of atmospheric carbon dioxide on the climate. Over the next two years the Jasons developed a climate model "which showed that doubling the carbon dioxide concentration of the atmosphere from its pre-industrial level (about 270 ppm) would result in an increase of average surface temperatures of 2.4°C".[20] This matched the 1977 findings of the National Oceanic and Atmospheric Administration.[21] When the Jasons report came in in 1979, all of a sudden—because it was the Jasons and had national security implications—this information and its devastating implications became important to the White House. However, President Carter (another Baptist) didn't last long; Reagan had other priorities, and a small group of scientists increasingly backed by fossil fuel lobbyists, got very busy.

Central to the post-1979 effort to discredit the science that showed anthropogenic climate change as a pending catastrophe were the economist Thomas Schelling and the physicist (and co-founder of the climate sceptic Marshall Institute) Bill Nierenberg. Magnifying the uncertainties in the data and modelling, the basic line of their argument was that "treating symptoms rather than causes would be less expensive, that new technology would solve the problems that might appear so long as the government didn't interfere, and that if technology didn't solve all the problems, we could just migrate".[22] But during the Regan Presidency (1981–1989) public awareness of climate science was mounting and 1988 was a year of terrible drought, so this resulted in the formation of the Intergovernmental Panel on Climate Change in 1989. Under the presidency of George Bush Sr., it looked like the United States was on track to do something serious about anthropogenic climate change.

Yet, several things were now working against a greening of US policy stances in the White House. Through the 1990s alliances between fossil fuel lobby groups, Republican politicians, and climate sceptic think tanks were shoring up, and the US love of conspiracy and the fundamentalist suspicion of science as a Trojan horse for Progressive anti-Christian reforms provided fertile grounds for political leverage through the Religious Right. The climate of fear and the sense that the American Way was under attack after 9/11 gave further impetus to a swing in the White House away from strong climate change action. This latter dynamic links consumerism, gas-guzzling cars, and global corporate power with the American Way, which under no circumstances must be pushed down by terrorists or anyone else. Come the first decade of the twenty-first century, even though the Democrat Obama gets elected in 2009, the tide had turned away from the Greening trend that was never firmly established, even as the findings of climate science became ever clearer and more urgent.

The Christian Right climate denial movement that has established a strong grip on US Evangelical Christians in the second decade of the twenty-first century, had powerful external drivers in the form of wealthy, media-savvy, lobby smart, fossil fuel, and big development interests. Equally, powerful internal theological and cultural drivers were at play in the Christian Right. Even so, it would be wrong to think that the greening trend in US Evangelical circles evident from the 1990s to the late 2000s was not a genuine and theologically sincere movement. In tandem with these drivers, an active refusal to take climate change mitigation seriously—as distinct from just ignoring the problem—took hold of the White House towards the end of the Clinton administration. As Oreskes and Conway note:

> In July 1997, three months before the Kyoto Protocol was finalized, US senators Robert Byrd and Charles Hagel introduced a resolution blocking its adoption ... Scientifically, global warming was an established fact. Politically, global warming was dead.[23]

In the first decade of the twenty-first century, an energised Religious Right linked into the Republican party, and supported by fossil fuel lobby groups such as the Global Climate Coalition and the American Petroleum Institute, was fired up to assert power. In this context any Progressive initiative that could be framed as anti-American (anti-Christian)—Greenies being an easy target—could be relied on to produce voter support from the Religious Right. And so the "climate science is a hoax" hypothesis was skilfully promoted to rally Evangelicals to the Republican side of politics. In this context Gore supporting Evangelicals were hung out to dry.

6.3 Katharine Hayhoe—Texan Evangelical Climate Scientist

In 2009, Katharine Hayhoe and her husband, Pastor Andrew Farley, published *A Climate for Change—Global Warming Facts for Faith-Based Decisions.*[24]

Hayhoe was one of the recipients of the 2007 Nobel Prize, with Al Gore, in relation to her work for the IPCC. She is a highly qualified climate scientist who is also a theologically Conservative Texan Evangelical. Either she or her husband (it is not clear who from the book) is a young earth creationist. But the point they are making is that whether one is a young earth creationist or not, climate science showing that climate change is a real and present danger is solid and true, and Christians have a theological responsibility to address this problem with urgency.

Hayhoe stands politely but firmly opposed to the views of other American Evangelicals promoting climate change inaction. In the first decade of the 2000s, she had considerable opposition on just that front:

Ann Coulter: "God gave us the earth. We have dominion over the plants, the animals, the trees. God said, 'Earth is yours. Take it. Rape it. It's yours.'"[25]

106 *Theology and Climate Change*

James Inhofe: "With all of the hysteria, all of the fear, all of the phoney science, could it be that man-made global warming is the greatest hoax ever perpetrated on the American people? It sure sounds like it."[26]

James Inhofe: "God's still up there. The arrogance of people to think that we, human beings, would be able to change what He is doing in the climate is to me outrageous."[27]

John Hagee: "Mark it down, take it to heart, and comfort one another with these words. Doomsday is coming for the earth, and for individuals, but those who have trusted in Jesus will not be present on earth to witness the dire time of tribulation."[28]

Coulter is a prominent lawyer, Inhofe is a US senator, Hagee is a Texan mega-church pastor—all are influential white American Evangelical Christians. This is a powerful and by no means insignificant religio-political stance in American politics. In this context, the valiant Hayhoe seeks to put forward an Evangelical case for climate change action, taking climate science seriously, and employing an Evangelical stewardship theology of creation. But by 2009, the tide had turned against her.[29]

By 2009, the Evangelical opinion tide had turned against the Evangelical greening. Even so, her book is interesting in the type of theological arguments she does and does not make. It has no engagement with Lynn White's argument and shows no interest in thinking of theology as itself a central factor in the production of our ecological crisis. Her book's first concern is addressing climate science scepticism, as this is obviously Hayhoe's strong suit, but it does so in ways carefully couched not to press the hot buttons of her audience. She makes it plain that she is not a tree-hugging, hemp-wearing, solar-powered, earth-worshipping, vegan Greenie—she is a normal Texan Christian living in the modern consumer age. But she is trying to get American Evangelicals to see the big picture of what climate science is really telling us, and to get behind policy-directed change that will turn us away from heavy carbon dioxide emitting energy dependencies. The assumption is that proper scientific education, sensitive to the peculiar political and theological sensibilities of American Evangelicals, will be all that is needed to enlighten her fellow Christians to the need to take climate change seriously and to change their ways as regards environmentally sustainable natural resource use.

Hayhoe is careful not to tell anyone they *have* to respond to the climate crisis, for this violates the concept of individual autonomy and personal spirituality so central to modern Evangelicalism.[30] Voluntarist and nominalist modernity is assumed here. Equally, a modern scientific positivism is assumed by her. While Hayhoe shows as much sympathy to climate change sceptics as she can possibly find, she knows the science, she patiently explains that scepticism is not scientifically justified, and she is trying to win them over to responsible climate stewardship on Evangelical terms. That is, she argues that responsible care for the globe's climate will not wreck our economy, it is not an anti-development stance, it is not a pagan stance, and it is compatible with a Christian understanding of the proper stewardship of nature as given to us by God. But her hardest hitting point is the same concern that is clearly central to what Pope Francis later wrote; the impact of climate change on the world's poor is, and will increasingly be, catastrophic.

> Love God, love others, and remember the poor; this is the unwavering mandate of the early church more than two thousand years ago. And this is our solidly biblical mandate for caring about climate change today.[31]

This is an ethically serious, science-based, but carefully moderated request for Evangelicals to open their hearts to treating climate science seriously.

Significantly, even though a prominent Texan pastor has co-authored this book with Hayhoe, there is very little in the way of first-order theological thinking in this book. Indeed, it is strategically important that first-order thinking is *not* there in this book. There is no radical call to re-think our theology, to fundamentally change our ways, to re-configure the way we feel and think towards nature, to re-evaluate normal way-of-life expectations, or to seriously upset the structures of interest and wealth generation associated with powerful corporate interests in the fossil fuel sector. Hayhoe and Farley are being very sensitive to the underlying feeling of insecurity within the traditional American Evangelical community on this issue. As a result they are not asking for fundamental change, they are arguing for a gentle re-ordering of normality that keeps the underlying structures of the American Way in place. But whilst keeping the American Way in place, they do see a pressing ethical and practical need for that way to be incrementally re-calibrated in the interests of the global poor, and in the interests of the American Way's own future viability. Whilst Hayhoe and Farley are genuinely seeking to have climate change addressing policies put in place, they remain (unsurprisingly) profoundly

108 *Theology and Climate Change*

committed to technological instrumental modernity. Evangelicals are intimately entailed in the American Way, and in the global network of power and interest that the United States upholds. This global network and its set of life-world expectations extends far beyond the United States. After 2008, the Evangelical outlook in the United States has swung decisively towards climate change scepticism. In this context it is no surprise that in Australia an Evangelical Prime Minister is a strong supporter of protecting Australia's coal exporting sector.

The context of trying to speak about climate change to fundamentalist-inclined, Religious Right leaning Evangelicals, after the effective political rise of Inhofe's "hoax" stance in US politics, strongly shapes the way any argument of this nature can be framed. But outside of that context, an Evangelical stance on serious climate change prevention can be framed differently. For one last look at Evangelicals, we will now turn back to Australia.

6.4 Ian Hore-Lacy on Responsible Dominion

Ian Hore-Lacy is a senior advisor with the London-based World Nuclear Association and an Australian Evangelical Christian. Just before the US Evangelical greening crashed, in the first decade of the twenty-first century, he wrote *Responsible Dominion: A Christian Approach to Sustainable Development*.[32]

Hore-Lacy is very critical of what he describes as a romantic and pagan eco-spirituality.[33] He shares this critique of pagan and pantheist eco-spirituality with Schaeffer, yet Schaeffer himself is clearly continuous with a romantic understanding of nature which has its roots in the nineteenth century Christian conservation activism out of which the contemporary Green movement actually arises.[34] Hore-Lacy decisively distances himself from that romanticism, maintaining that "the modern neo-romanticism is an essentially religious attitude which largely drives parts of the environmental movement".[35] Like Schaeffer, and appealing to Alister McGrath,[36] he is very critical of Lynn White.[37] He upholds a responsible utilitarian approach to nature that recognises that God provides for us with nature's bountiful resources, and thus Hore-Lacy is sceptical of the notion of natural resource exploitation limits.[38] He is an Evangelical and, like Denis Alexander,[39] sharply critical of what he sees as the shameful anti-science baloney of young earth creationists.[40] He is critical of the influence of the fantasy writings of C.S. Lewis and J.R.R. Tolkien as they give a populist Christian anointing to what he sees as pagan romanticism.[41] Hard science and a wise use of instrumental technological rationality really matters to Hore-Lacy, who admiringly quotes another Evangelical Christian, Sir John Houghton, recognising the problem of climate change as needing urgent attention. Hore-Lacy sees nuclear energy as the obvious answer to climate change, with renewable energy sources providing only a small contribution to our energy needs "at the margins".[42] In no manner is he denying climate change, but he has as firmly an instrumental and non-sacral approach to nature as any mining company director. Indeed, Hore-Lacy sees a solidly modern, scientific, and instrumental approach to creation as incumbent upon Christians in order to lift the global poor and feed and prosper all the citizens of the globe. There can be no doubt that he would find Saint Francis' creation spirituality, Orthodox sacramental conceptions of nature, and Celtic and Indigenous Christian theologies of nature to be tosh at best, and most probably heretical folly.

The key theological concept to Hore-Lacy is the Evangelical theologian John Stott's term, "responsible dominion."[43] Hore-Lacy refers to Stott when describing the responsibility of Christians to take environmental issues seriously, but in a distinctly Christian manner.[44] What Stott means by a biblical Christian understanding of dominion is a dual ownership idea.[45] That is, as belonging to the Creator, "The earth is the Lord's and everything in it" (Psalm 24:1). Yet God has given man dominion over the earth (Genesis 1:26), so it is also true that "the highest heavens belong to the Lord, but the earth he has given to man" (Psalm 115:16). Nature is given to us to rule and use, but as a tenant who is held responsible for how he uses it, rather than as a final owner. A tacit separation between the realm of God (supernatural heaven) and the realm of man (the natural earth) maps neatly onto the separated domains of personal faith and public action native to a modern idea of how religion itself functions within a Liberal secular democracy. It is also clear from Stott that technological exploitations of nature are to be celebrated as a vital feature of man's proper dominion over the earth. Stott notes:

> Developing tools and technology, farming the land, digging for minerals, extracting fuels, damming rivers for hydro-electric power, harnessing atomic energy – all are fulfilments of God's primeval command. God has provided in the earth all the resources ... we need, and he has given us dominion over the earth in which these resources have been stored.[46]

We fulfil what God requires of us in relation to creation when we recognise that "our dominion over the earth has been delegated to us by God, with a view to our co-operating with him and sharing its produce with others; then we are accountable to him for our stewardship."[47] But—acknowledging God or not—humanity has a real dominion over nature, for human benefit, and nature is stocked abundantly with everything we could ever need, so there is no shortage in what God has provided for us. And in itself, there is no sacred value (though creation has use value, certainly) in the earth; Christians are not pagans.

The thing I would like you to notice here is that this sort of Christian approach to nature is completely modern, 100% at home in the scientific and technological age, and, at least functionally, pragmatic and rationalistic with regards to how a responsible use of nature should be pursued. This stance denounces "romance", "ideology", and "pagan" thinking, and renounces superstition, pseudo-science, and any form of epistemic or use taboo. A secular materialist who wants to pursue a rational utilitarian and technological approach to the common good could, in principle, be entirely functionally at one with such a stance. And this is the point I would like you to notice. Ian Hore-Lacy has an explicitly Christian theology for his functionally materialist and pragmatically instrumental understanding of what a responsible and sustainable use of the natural resources of the planet should look like. You could pursue exactly the same line of reasoning and practice, if you had pragmatic reasons to want to address climate change and a humanitarian interest in sustainable development in the global south, without his Christian theology, and without any religious commitments at all. Perhaps this might be the case because pragmatic materialists have embraced a tacitly Evangelical theology of nature?

6.5 Evangelicals and Lynn White

By this stage it may well be clear that many Evangelicals—like many agnostic Conservatives and atheist Progressives—take climate science and global warming very seriously. Further, even though many Evangelicals have specific theological reasons for not wanting to be affiliated with the Green movement, Evangelicals (including young earth fundamentalists) remain remarkably modern, Western, and instrumental in their basic approach to nature, just like many agnostic and atheist modern Western people are. This indicates to me that Progressive Dominion Theology is no longer a uniquely religious (i.e. Christian) stance towards nature, but it is a broadly inclusive modern, Western, and instrumental stance towards nature. But such a stance—as Lynn White perceived—does genuinely have a Christian theological genealogy. As contemporary modernity has its roots in the Christian world of seventeenth-century Western Europe, Evangelicals may tell us more about the basic first-order beliefs and attitudes to nature in secular modernity than anyone else does. The relationship between the conscious Progressive Dominion Theology of Evangelical Christians and the unconscious Progressive Dominion Theology of secular pragmatic functional materialist, is thus a very illuminating relationship. This relationship can uncover the theological reasons at the core of why, culturally, we find climate change such a difficult reality to address with appropriate seriousness and urgency, for it is our theology that provides the real grounds of resistance to any first-order re-structuring of our very way of life. Lynn White was right; a distinctive form of Western theology, now deeply embedded in the very fibre of our common way of life, is intimately causally entailed in climate change. And we can get to that theology by looking at where it is closest to the surface—Evangelical Christianity. But note carefully, this observation is in no manner intended as an anti-Evangelical observation. As a disclosure, I would not be interested in saying this if I did not think both Evangelical theology and Western modernity could not be transformed. I think a careful understanding of Christian theology provides us with an answer to our present ecological crisis; I do not see theology as a tool for asserting shame and blame.

There are two further things to add here. Firstly, atheistic materialism arises easily from a theological view of nature that is entirely desacralised, that has a religiously unrestrained pragmatic and instrumental approach to nature, and where the supernatural is entirely discrete from the natural. This is justified by a distinctive theological perspective to start with, but once an entirely naturalistic understanding of nature is theologically established, and once a pragmatic imprimatur in the use and exploration of nature is also theologically established, the practice of instrumental naturalism no longer needs a theoretical underpinning in Christian theology. The relationship between Christian theology that is native to modernity and that gave rise to modern science, and to agnostic or atheistic scientific naturalism, is intimate. The relationship between an instrumental natural domination theology (be it responsible or not) and an entirely pragmatic and materialist exercise of self-interested power over nature, is intimate. The CEO of a coal mine could be as at peace with herself in exercising responsible dominion over nature as an Evangelical Christian. Equally, a non-religious CEO of a coal mine could have a good conscience in performing a valuable job for her corporation exploiting a profitable natural resource. If one is a pragmatist in practice, then the successful exercise of power and the valid pursuit of enlightened self-interest—to the utility of one's self, one's company, and those who benefit from one's service—is its own reward. If the resource is there and its extraction and sale economically viable, someone has to profit from it; why not me?

Secondly, PDT is a *political* theology. Dominion is about power. This is by no means a reductive statement. Power is a very complex animal, and power actually requires a theology of one sort or another. We will now look briefly at PDT as a political theology.

Notes

1 The main politically relevant sub-set of "Evangelical" this chapter is concerned with is the American Evangelicals that Veldman describes as "traditional Evangelicals" who are now usually referred to as "white Evangelicals" in the broader literature. This racial qualifier is significant and cannot be avoided.
2 This is complex as, for example, the Clapham Sect was the main driver behind anti-slavery parliamentary reform, and Methodists were often involved in nineteenth-century labour reform and other Progressive endeavours for the rights of women and children. There is a strong history of Evangelical left political inclination as well. In this context it is very important to note that whilst the Religious Right in the United States is largely an Evangelical movement, being an Evangelical does *not* necessarily equate with supporting the Religious Right.
3 This is the famous legal struggle in 1925 concerning teaching of evolution in US public schools.
4 Evangelical disagreement with Darwin was not radical opposition in the nineteenth century, as seen in Samuel Wilberforce's detailed interest in Darwin's work.
5 Veldman, *The Gospel of Climate Skepticism*.
6 Schaeffer, *Pollution and the Death of Man*.
7 See for example, Guinness, *The Dust of Death* [1971].
8 Though, it was the Republican administration of Richard Nixon that set up the Environmental Protection Agency in 1970, the same year Schaeffer published his book on pollution.
9 In this regard Schaeffer runs remarkably parallel with the recovery of patristic and medieval theological themes in the work of thinkers like C.S. Lewis in the mid-twentieth century and Catherine Pickstock in the present. Schaeffer held that moral issues *are* aesthetic issues, and vice versa. See Pickstock's introduction to Gill, *Beauty Looks After Herself*. As Michael Northcott points out, the nineteenth century conservation movement is deeply embedded in a Christian Romantic sensibility. This is part of what Northcott describes as the "pro-ecological turn of Protestantism." See Northcott, "Lynn White Jr. Right and Wrong."
10 See Butterfield, *Christianity and History*, 9–18.
11 Schaeffer, *How Shall We Then Live?*

12 Veldman, *The Gospel of Climate Skepticism*, 132–138, 198–199. Veldman documents how the Ethics and Religious Liberty Commission of the Southern Baptist Convention – which she describes as "the Southern Baptist Convention's powerful political arm" (135) – set about directly opposing the stance Jonathan Merritt put forward. See the document Merritt initiated: "A Southern Baptist Declaration on the Environment and Climate Change," reproduced in Veldman, 229–233.
13 That is, Schaeffer is more concerned about the anti-Christian *reception* of White's argument in the context of the conservation movement of the 1970s, than he is about the argument itself. For a very fine piece looking at reception and interpretation issues flowing from White's argument, see Harrison, "Subduing the Earth."
14 See, for example, Alexander, *Creation or Evolution: Do We Have to Choose?*
15 Merritt, *Green Like God*.
16 Veldman, *Gospel of Climate Change*, 5. Citations for this paragraph are all detailed in the notes to this page in Veldman's book.
17 Veldman, *Gospel of Climate Change*, 236–237.
18 Oreskes and Conway, *Merchants of Doubt*, 172.
19 Finkbeier, *The Jasons*.
20 Oreskes and Conway, *Merchants of Doubt*, 171.
21 As cited in Oreskes and Conway, *Merchants of Doubt*, 172; see White, "Oceans and Climate – Introduction," 2–3.
22 Oreskes and Conway, *Merchants of Doubt*, 183.
23 Oreskes and Conway, *Merchants of Doubt*, 215.
24 Hayhoe, *A Climate for Change*.
25 On the US TV show, Hannity and Colmes, 20 June 2001.
26 James Inhofe US Senate floor speech, "Catastrophic Global Warming Alarmism Not Based on Objective Science" delivered on July 28, 2003. Above citations are found here: https://www.nytimes.com/2003/08/05/science/politics-reasserts-itself-in-the-debate-over-climate-change-and-its-hazards.html?pagewanted=all&src=pm. See also Inhofe, *The Greatest Hoax: How the Global Warming Conspiracy Threatens Your Future*.
27 As cited in Veldman, *The Gospel of Climate Skepticism*, 6.
28 As cited in Bill Moyers, "Welcome to Doomsday," https://billmoyers.com/2005/02/25/welcome-to-doomsday-march-24-2005/.
29 Veldman, *The Gospel of Climate Skepticism*, 161–189.
30 Hayhoe, *A Climate for Change*, 129: "Seeing the effects climate change is having on our world may leave us cowering in fear or overwhelmed with guilt. But God is not the author of fear or confusion ... He does not use guilt to motivate us." Hayhoe, *A Climate for Change*, 139: "If you decide you don't want to individually contribute to a solution to climate change, so be it. You are free in Christ to decide that."
31 Hayhoe, *A Climate for Change*, 127.
32 Ian Hore-Lacy, *Responsible Dominion*.
33 Hore-Lacy, *Responsible Dominion*, 30–31.
34 Again, see Northcott, "Lynn White Jr. Right and Wrong."
35 Hore-Lacy, *Responsible Dominion*, 31.
36 Hore-Lacy maintains that Alister McGrath has provided a knock-down argument against White in McGrath, *The Re-enchantment of Nature*.

37 Hore-Lacy, *Responsible Dominion*, 24–25.
38 Hore-Lacy, *Responsible Dominion*, 23, 27.
39 See, for example, Alexander, *Creation or Evolution: Do we have to Choose?* On page 36 of *Responsible Dominion*, Hore-Lacy cites Alexander, *Rebuilding the Matrix* as the best paradigm for a Christian to understand science-faith issues.
40 Hore-Lacy, *Responsible Dominion*, 34.
41 Hore-Lacy, *Responsible Dominion*, 45: "Green Romanticism, even if anointed by Christians, is a cop out." See 133–134 for his identification of Lewis and Tolkien with populist Green Romanticism.
42 Hore-Lacy, *Responsible Dominion*, 93. See also this short piece by Ian Hore-Lacy and Robert Farago's response. Hore-Lacy, "Australia's Energy Insanity," 20 July 2018, *Ethos*. Farago, "Nuclear Has Left Its Run Too Late: A Response to Ian Hore-Lacy," 14 August 2018, *Ethos*.
43 Stott, *Issues Facing Christians Today*, 114.
44 Hore-Lacy, *Responsible Dominion*, 16.
45 Stott, *Issues Facing Christians Today*, 109–121.
46 Stott, *Issues Facing Christians Today*, 113.
47 Stott, *Issues Facing Christians Today*, 115.

7 Climate Change and Political Theology

In relation to the thought categories used in this book, there are two types of political theology: religious political theology (Political Theology B) and philosophical political theology (Political Theology A).

7.1 Religious Political Theology and Climate Change

Religious political theology is explicitly religious theology used in a political way. The relationship between the Religious Right, the Republican Party, and climate policy in the United States is a potent, indeed world-affecting example of this. Robin Globus Veldman's exceptionally fine look at that relationship should be read closely by anyone seeking to properly understand this globally significant example of religious political theology.[1] One important political aspect of this dynamic concerns end-time thinking.

7.1.1 Politics and Apocalypse

Apocalyptic consciousness is a recurring feature of Western culture. When a sense of pending existential threat arises to the norms of the prevailing way of life, the apocalyptic tropes deeply embedded in the Christian roots of Western modernity come to the fore. This can come to the fore in secularised forms, such as in the popularity of dystopian and post-apocalypse movies, as easily as in explicitly religious forms. The threat of nuclear winter during the Cold War contributed greatly to the rise of apocalyptic thinking in US Evangelical circles, which has been thriving from the latter half of the twentieth century to the present.

Politically, we can also see that it is in times of radical future uncertainty that demagogues arise offering an "Other" to blame for our insecurity, and offering decisive externalised solutions to our deepest fears that don't require any re-ordering of the priorities and commitments of our own (threatened) way of life. There are known political dynamics that go with rising levels of apocalyptic consciousness. In this context, it is important to note that the real driver of climate change scepticism in Evangelicals is not a function of their views on science. It is an underlying sense of insecurity that the very way of life they are embedded in is under threat. And on that point, we are all in the same boat. Evangelicals frame their sense of being threatened in a particular way, and this way needs to be properly understood if the political dynamics of Evangelical climate change denial are to be adequately understood.

7.1.2 The Embattled Evangelical

As our brief exploration of Schaeffer has shown, it is first-order framed theological concerns and their cultural and political implications that most fundamentally concerns traditional Evangelicals. Once they became politically activated as an electorally powerful interest group in the 1980s, they did not want to make political alliances with ideological Progressives (functional materialists and socialists), or theological Liberals (heretics), or identity interest groups who they see as attacking the American Christian Way (gays, Muslims, feminists, etc.). Politically speaking, there is a sound logic here. If, for example, an Australian Evangelical wants to vote for the Green party because she supports climate change mitigation, and because she finds the "border sovereignty" logic of the main parties unconscionably inhumane towards boat arrival refugees (two classically recognisable Evangelical political concerns), she is also voting for abortion, gay rights, and the de-funding of Christian schools. So even though the environmental conservation movements of the nineteenth century were largely driven by Christians, the Green movement is now largely seen as lost to the Evangelical *political* cause. Evangelicals tend—often accurately—to view the post-war Greens as some combination of materialists, Liberals, pantheists, and animists in their primary commitments, which certainly does shape their sense of political mission.

Once Evangelicals are a defined political interest group, they then have the problem of not being yoked to political interest groups that are directly opposed to their own first-order commitments. Interestingly, pro-business, pro-industry, pro-capitalist political interests that are also culturally Conservative—notably, the Republican Party in the United States—are seen by many Evangelicals as politically compatible. This really is interesting as it suggests that indeed there is a deep commonality between the basic theological commitments of secular wealth and power in the United States, and Christian Evangelicals. This alliance suggests that Progressive Dominion Theology is the common ground between the Republican Party and traditional American Evangelicals.

Modern North America has a Puritan origin, birthed in the apocalyptic turmoil of seventeenth-century England. The Evangelicals, being dis-established from the outset, have thrived since the American Revolution, and are deeply embedded in the American culture of God, freedom, industry, and enterprise. In the nineteenth century, Evangelicals were very prominent in public life, and—without there being an Evangelical political voting block as such—they were politically powerful.[2] Since the late nineteenth century, they have felt their power waning and they have felt that the commitments that make the United States a Christian nation have been relentlessly eroded by the rise of adverse cultural forces. All this is true. There is nothing made up about their feeling embattled and sensing that their culture shaping and political influence has been eroded. These are the forces that have motivated the rise of the Religious Right and that have defined the sort of political alliances they have made.

Veldman also explores the manner in which, once the Moral Majority started, Evangelical leaders have made careful alliances with wealthy Republican Party and business partners, and have harnessed the enormous Christian mass media in the United States to shape the attitudes and identity markers of the larger Evangelical population of the United States.[3] There is nothing accidental about the rise of what Veldman calls "the gospel of climate scepticism" in twenty-first-century American Evangelicals. And clearly, the Political Right needs the Evangelical voter base just as much as Evangelicals need access to political influence. So the relationship is strong, but it is not simply pragmatic. It is undergirded by a common tacit theology. And that theology is basic to The American Way, as understood by cultural Conservatives in both Evangelical and secular America, particularly in business and finance. Yet, an embattled cultural identity politics does not only define Conservatives, it defines Progressives as well.

7.1.3 The Embattled Environmentalist/Progressive

Any community that feels its way of life is under threat by large and insidious cultural forces will develop an embattled mindset. This is just as true for environmental activists who identify as Progressive as it is for traditional Evangelicals.

Susan George, in *Hijacking America*, makes an impassioned and yet carefully documented analysis of the decisive shift to the right in US politics that came to maturity in the first decade of the twenty-first century.[4] George sees the political, media, and wealthy lobbying alliances forged between the Religious Right and the George W. Bush administration as a concerted assault on the Progressive Enlightenment-framed commitments of modern America.[5] To George, the Enlightened, scientific, and secular America—and even its Mainline religion—that she believes in, is under savage assault. George documents the way the Progressive moral values that she (born in 1934) grew up with were being thrown under the bus in the first decade of the twenty-first century. The commitment to scientific truth was being deeply eroded by religious fanatics[6]; the commitment to the rights of all individuals to equality and freedom of belief and expression was being hijacked; and the emancipation of women, blacks, gays, and other minorities from a repressive cultural norm and power inequalities was being wound back by Conservative (white, gun-owning, bible-believing) Christian prejudices and interests. The way political influence was bought by wealthy lobby groups is carefully documented by George, and this is something she finds particularly egregious for liberal and democratic ideological reasons. George sees the neoConservative ideological shift away from the sovereignty of the people to the sovereignty of markets, which gained political traction under President Reagan, as radically transitioning Enlightened America away from its founding political roots.

George closely maps the "long march through the institutions"[7] to cultural hegemony achieved over time by deep ties forged between the Christian Right, Republican Party politics, and wealthy lobby groups. Writing before the election of President Obama, George presciently saw that the shift to the right was far deeper than could be halted by a Democrat President. The election of President Trump in 2016 and the ineffectiveness of the Obama administration in re-regulating the finance sector after the 2008 financial sector crisis rather supports her forward analysis.

The culture wars in the United States are fought between two now deeply antagonistic sides: Embattled Conservatives and Embattled Progressives. Twentieth-century Progressives are strongly tied to the Conflict Thesis where "religion" is the enemy of Enlightenment. Thus, Young Earth Creationism is an ideological lightning rod supporting the Progressive account of irrationality and anti-science backwardness as the central defining features of the Religious Right. Progressives see the cultural influence of the Religious Right as an assault on modern America, powered by the heady mix of religiously induced false consciousness linked with very wealthy private interests, the Republican Party political machinery, and sophisticated lobby and media networks of serious mass persuasion force.[8] I have outlined how Conservatives feel embattled and why, but it is certainly the case that Progressives who had such high hopes for progress in the twentieth century are feeling savagely disempowered and assaulted by the Religious Right.

The difficulty and need for a genuine and first-order response to climate change is particularly felt by the young Progressives who are inheriting a world spiralling into man-made catastrophic uncertainty on an unprecedented global scale. The Thunberg/Ernman family's profoundly frustrated anger about the failure of climate change mitigation action is grounded in factual evidence, undeniable prudence, and a deep sense of the moral failure of the prevailing status quo.[9] The inertia against concerted climate change mitigation is indeed—as Susan George documents—promoted by powerful financial, religious, and political Conservatives.

Progressives are on the back foot as the Conservative forces of climate change inaction have gained an upper hand in the politics of the United States. As the global leader of the post-war world order that the United States largely constructed, this Progressive sense of embattlement is entirely justified. Embattlement in what is seen by both sides as very high stakes, characterises the toxic culture war in the United States. The future of the global eco-system is one of the policy footballs in this power struggle.

7.1.4 The Politics of "Out-Groups" and a Kerygmatic Note on Political Strategy

Veldman accurately identifies the known socio-political dynamic of identity-based out-grouping as a key component of climate change politics. Veldman explains:

> ... Evangelicals framed environmentalists as an out-group [and] .... environmentalists also view Evangelicals as an out-group.[10]

Both groups define their own identity as not being like the enemy out-group. For this reason, it is very hard for any member of either identity group to "reach out" to the other side and find some sort of common ground.

As pointed out in the introduction of this book, I think climate change requires a concerted global response that is unflinching in its preparedness to look at first-order life-world outlooks and practical commitments. I also think that US Evangelicals, as a voting bloc, are the key to producing political change on a global policy level. So, in this short section I am going to depart from an objectively voiced academic genre and lean, momentarily, into a kerygmatic genre to ask, how might Evangelicals be converted away from the gospel of climate scepticism?

Recognising that the culture war between Conservatives and Progressives frames the political realities of this question, both sides need to understand each other better—and not just as enemies—for change to happen in this field. Let us explore the basic "reach out" obstacles to both sides, one at a time.

Progressives seeking to "reach out" to Conservative Evangelicals on the matter of climate change will make no working contact if they treat Evangelicals like pre-scientific nutjobs and if they expect to make political alliances with Conservatives on most other classically Progressive social reform concerns. It must be recognised that the reason science itself is not the first point of contact in this context is because—to the traditional Evangelical—deeper cultural drivers are at play. The meaning of the cosmos, the presumed anthropology, the assumed philosophy of nature, and the way of understanding environmental issues through a creation care lens will not be compatible with belief and action categories that are amenable to many modern Progressive first-order assumptions. If Progressives can respectfully gain access to the Evangelical mind—without, of course, pretending to share the same commitments—then science can be raised. Evangelicals are actually modernists, they have no genuine connection to pre-modern ways of thinking, and scientific evidence is genuinely important to them. The evidence for climate change is measurable in the present in relation to the immediate past; this is a different sort of science question to the origin of the human race. Young Earth Creationism need not come into the science of climate change. I think trying to isolate climate change from the larger culture war struggles between Progressives and Conservatives is the best strategy here. If a limited alliance can be forged, just on climate change action, that would be a success that may well have political viability.

Conservative Evangelicals seeking to "reach out" to Progressive environmentalists need to stop treating Greenies like the devil incarnate. They may indeed be materialists or pantheists, they may indeed find animism interesting, and they may indeed be moral and cultural constructivists who are pro-gay, pro-abortion, and broadly anti-Christian. But if Christians have a responsibility to properly steward the earth such that the global poor do not suffer catastrophic loss through climate change, so that the resources of the earth can indeed be sustainably shared with justice for all the people and creatures of the earth, and so that God's beautiful creation is properly protected and nurtured, then Evangelicals need to work out how to find common political cause with environmentalists on serious climate change mitigation action. This will have costs for fossil fuel industries. This will have lifestyle implications for Americans. This will require discarding the prevailing ideology of perpetual economic growth.[11] Evangelicals cannot be beholden to financial, commercial, and lifestyle status quo power blocs, if that status quo is prepared to resist threats to their wealth and power at any cost. There is some serious soul searching that needs to be done in relation to God and Green that was systematically squashed when Jonathan Merritt tried it, and that squashing needs to be owned as a mistake and repented of. There are signs that this side of Evangelical thinking is still there and could be revived.[12]

Enough preaching about political pragmatics. Now to a very quick look at first-order political theology concerns.

7.2 Philosophical Political Theology and Climate Change

In the West's political history, power has been intimately theological from the Hebraic origins of divine command thinking, from the central role of cultus and religious festivals in the Greek City State, through the Greco-Roman imperial era, through Constantine, through the "dark ages," medieval Christendom and the great age of popes and kings, through the rise of the modern nation-state, and through to the recent emergence of liberal, secular, and democratic modernity.[13] The idea that politics is its own thing, fully autonomous from theology, is a very recent idea. Such an idea is *not* there in Pufendorf or Grotius in the seventeenth century, God *is* there in the eighteenth-century US Declaration of Independence, and the Crown—deriving its authority from God—*is* there in the twentieth century Australian constitutions as the final grounds of political authority. Indeed, to this day, the English monarch, as the final authority of the Australian state, is also the head of the Church of England; in theory at least, secular, multicultural, post-Christian Australia still has something resembling a structurally established church. Even post-modern power configurations, through various forms of spectacle and personal identity construction, are theological.[14]

In modern political history, institutional separations of the Church from the State do not necessarily imply the autonomy of political authority from theology. And yet, the temporally localised influence of reductive naturalism and secularisation over the past 150 years, and the concerted effort of Whig reformers from the late nineteenth century on in the English legal system, makes political theology seem like a strange idea to us. But, strange or not, it is still there. In a manner quite consistent with our earlier discussion of Aristotle's understanding of first philosophy (Theology A), the final grounds of political authority have always been, and remains, "divine". That is, the notion that political and legal authority is equivalent with either mere force or the arbitrary (merely procedural) self-determination of the people is a rejection of the Western notion of political authority; *just* (as in, "of justice") power is the ground of true political and legal authority. Just power is not something that force or votes can magically produce.

Michael Northcott has written an astonishingly philosophically powerful examination of the contemporary relationship between political theology and climate change, which I recommend to anyone seeking a serious understanding of this vitally important area.[15] I will not seek to summarise that text here, but it is a very helpful door opener as to why we find it so hard to respond to climate change. In recent decades, the undergirding assumptions of valid power have become operationally embedded in borderless trade, virtual finance, the pragmatics of electoral success, consumer culture norms, technological and individual freedom, and the dependence of the modern state on certain forms of mass-media persuasion. In consequence, we have come to accept that arbitrary power and the mere persuasion of a voting public magically generates political authority. For this reason, our politics is becoming increasingly sub-human, increasingly about the mere necessities of the status quo, mass persuasion, and force. Politics is increasingly the plaything of short-term, arbitrary, lowest common denominator, mass-media manipulation (the public opinion bulls and bears of fear and greed). For these reasons, our politics is increasingly unable to respond to a first-order global crisis like climate change. And, as Northcott points out, it is the tacit theological assumptions about personal freedom, the "logic" of sovereign markets, and the meaning of humanity and nature all embedded in our unexamined shared political theology, which makes it so hard for us to reform a mode of life that is inherently destructive of the natural world. As long as we don't think about the role political theology plays in providing us with the first-order structuring reality of the present norms of commercial, financial, and political power, we will remain the playthings of the entrenched political theology that is driving us to destruction. Northcott's book makes it clear that there can be no genuine or effective re-think about the life-world norms ordering our inherently environmentally destructive mode of life without first recognising the theological framework that orders our assumed power norms.

Underneath the culture wars that are defined by religious political theology, there are far deeper philosophical political theology frameworks of assumed meanings and entrenched habits defining our common way of life. Ironically, there is not a lot that separates the Progressives from the Conservatives when it comes to those underlying commonalities. As we have outlined in the first half of this book, the cultural structures of Progressive Dominion Theology run very deep and are embedded in how all of us now practice knowledge and power, and how we all understand nature and humanity.

7.3 Theology, Climate Change, and Politics: Concluding Thoughts

In this short book, I hope to have shown that many pious Conservative Evangelicals and many hard-headed business Conservatives are both deep natives of Western scientific and technological modernity. This is why "religion" is unimportant in understanding what they have in common. Surprisingly, it is theology, rather than religion, that they have in common. They share an instrumental Progressive theology of nature premised on the assumption of human dominion over nature. Science as a prophetic voice signalling the end of the modern world—even though it is science and technology that produced the modern world—is simply unwelcome to people totally committed to the prevailing structures of social reality. Unsurprisingly, those who most benefit from the modern world are most uncompromisingly committed to its continuation—these are not often scholarly or religious types, but are often business, politics, and finance types. Science and technology as a tool of PDT, and as functionally subservient to progress and power, is science as PDT and the functional norms of the modern Western way of life requires it to be. Science, here, is the servant, not the master, of the modern world.

PDT is not just for Conservatives, pragmatists, and the necessities that govern wealth and business. PDT is the tacit theology of all Western modernists: Progressives, the irreligious, post-truth constructivists, consumerists, and scientific atheists. It defines the imaginative horizons and the practical norms of our distinctive way of life. It defines our values.

PDT values power, it values domination, it values instrumental and calculative reason that solves problems, it values nature as a resource. We would not have Western modernity as we know it without this value system. And it is obvious that to question how our society approaches power and nature and human success, is to question everything about our very way of life. This is a very unsettling thing to do and Greenies know as little as anyone else about how seriously reframing our very way of life might be anything other than a huge step backwards in terms of the power and privilege we now see as normal, and as the right of any successful modern Westerner. We find it almost impossible to imagine, for example, a balanced and prosperous economy that does not require economic growth.

PDT has its collective psychological signatures too. The unwelcome side effects of the structural norms that we live under entail a certain degree of fatalistic shame that we find very hard to deal with. Homelessness brings this dynamic out.[16] When asked for some money by a rough sleeper, I notice in myself a degree of shame just because I am not living on the street, and the problem that causes this situation is obviously unaddressed by me giving a few dollars. I would rather just not see the problem. When many thousands of people died and millions were made homeless in Manila by typhoon Haiyan—which was indeed a symptom of global warming—we simply don't want to know about it.[17] The problem seems too big, our capacity to make a difference too small, and the guilt and shame of living in relative safety too hard to address. We would simply rather not know. This not wanting to know is far more significant than we give it credit for when it comes to climate change scepticism. And it really is naive, as the Noble Prize-winning economist and psychologist Daniel Kahneman points out, to think that such "non-rational" factors are not profoundly entailed in human decision-making and behaviour. Kahneman describes climate change as the perfect storm where ignoring the problem is by far the most likely (and psychologically understandable) human response.[18] Reductively scientific facts are but one feature in real human decision-making. Theology and psychology are inescapably entailed in human choices, typically at levels far beneath the "rational" surface of our decision-making. Interestedly weighted prudential judgements—again, far beneath the surface of our rationalized decision making justifications—are inescapably entailed in human choices. And, as Kahneman points out, if you can get preachers—people who work in the realm of theology and worldview entailed persuasive reasoning—to grasp the need for action, this is the kind of approach that just might work in *real* human contexts. But such insight is hard for our academic culture to embrace, particularly when it comes to thinking about religion.

In reading through an excellent scholarly book trying to understand why fundamentalism hadn't simply died out, the renowned sociologist Peter Berger found himself wondering

> *Who* finds [the world of fundamentalism] strange? Well, the answer to *that* question was easy … professors at elite American universities. And with this came the *aha!* experience. The concern that must have led to this Project was based on an upside-down perception of the world, according to which "fundamentalism" … is a rare, hard-to-explain thing. But a look at either history or the contemporary world reveals that what is rare is not the phenomenon itself but the knowledge of it. The difficult-to-understand phenomenon is not Iranian mullahs but American university professors … My point is that the assumption that we live in a secularized world is false. The world today, with some exceptions… is as furiously religious as it ever was, and in some places more so than ever. This means that a whole body of literature by historians and social scientists loosely labelled "secularization theory" [where Modernization necessarily leads to the decline of religion] is essentially mistaken.[19]

Religion and theology are not anomalies in the world today. But if our academic culture cannot take it seriously, we will not respond to its political implications intelligently.

With various caveats and qualifications, it seems undeniable that Lynn White is basically right. The shared theology of our deepest attitudes to nature, humanity, and instrumental power is what shapes the primary cultural groves and the entrenched structures of power that define normality to the now globally influential Western way of being in the world. The origin of that outlook is Western Christian theology. And indeed, PDT has served us well in producing the wonders and powers of the modern scientific age, but we have come to a point where it will prove our downfall if we cannot radically alter it.

With Lynn White, I am inclined to think that the religious source of the undergirding theology of Western modernity is the most obvious place to turn to as a starting point in re-configuring our theology of Nature and our theology of Humanity. As we saw in Chapter 5, there are very promising attempts to do just that in global Christian circles. Yet, these attempts are still hampered by the recent construction of the politically isolated and deeply individualistic conception of "religion" which, unsurprisingly, is itself a function of PDT. It is also the case that the most decisively modern, entrepreneurial, individualistic, and secular compatible form of Christian faith ever seen—white American Evangelical Christianity—is central to the current politics of climate change. Even so, a shift in the theology of nature here would be the decisive tipping point for the rest of the United States, and for the modern world.

The good news is that Evangelicals *are* interested in theology, and they *are* interested in Christian theology. The bad news is, if the Evangelical theology of nature cannot be isolated from the culture war between Conservatives and Progressives then we are probably doomed to unfixable climate catastrophe.

Theology is central to the real politics and the real life-world dynamics of climate change. A lot is going on in this field, but the time horizons for genuine change are short. In this context, clarity in understanding the very real impact of Christian theology on secular, Western modernity, is a vital ingredient in trying to address some of the most powerful cultural drivers of anthropogenic climate change.

Notes

1 Veldman, *The Gospel of Climate Skepticism*.
2 Noll, *America's God*.
3 Veldman, *The Gospel of Climate Skepticism*, 161–214.
4 George, *Hijacking America*.
5 George, *Hijacking America*, 185: "... the assault on the Enlightenment is in full swing."
6 George, *Hijacking America*, 160: "The religious program is progressing so fast, in fact, that the apparent triumph of science, even in the West, is under threat and may be far more fragile than we might assume."
7 This is Susan George citing Antonio Gramsci in *Hijacking America*, 100.
8 For a careful exposition of why embattled Progressives see the relationship between the Republican Party and the Religious Right as profoundly disturbing and dangerous, see Posner, *Unholy – Why White Evangelicals Worship at the Alter of Donald Trump*.
9 Thunberg, *Our House Is on Fire*.
10 Veldman, *The Gospel of Climate Skepticism*, 219.
11 Jackson, *Prosperity Without Growth*.
12 Subramanian, "Generation Climate: Can Young Evangelicals Change the Climate Debate?"; Subramanian, *Finding Middle Ground*. See also these US Evangelical organizations, Young Evangelicals for Climate Action: https://yecaction.org/; Evangelical Environmental Network: https://creationcare.org/. See the National Association of Evangelicals 2015 statement on climate change "Loving the Least of These. Addressing a Changing Environment": http://nae.net/wp-content/uploads/2015/06/Loving-the-Least-of-These.pdf.

13 On this last point, see Hunter and Guinness, *Articles of Faith, Articles of Peace.*
14 Ross, *Gifts Glittering and Poisoned.*
15 Northcott, *A Political Theology of Climate Change.*
16 This is explored with astonishing simplicity and depth in the children's story by Maggie Hutchings, *I Saw Pete and Pete Saw Me.*
17 Vidal and Carrington, "Is Climate Change to Blame for Typhoon Haiyan?"
18 Kahneman, "Daniel Kahneman on Misery, Memory, and Our Understanding of the Mind."
19 Berger, *Desecularization*, 2.

Bibliography

Aechtner, Thomas. "Galileo Still Goes to Jail: Conflict Model Persistence within Introductory Anthropology Materials." *Zygon: Journal of Religion and Science*, Vol. 50, No. 1 (March 2015), 209–226.
Agamben, Giorgio. *Homo Sacer: Sovereign Power and Bare Life.*, trans. Daniel Heller-Roazen. Stanford: Stanford University Press, 1998.
Albeck-Ripka, Livia., Jamie Tarabay, and Isabella Kwai. "As Fires Rage, Australia Sees Its Leader as Missing in Action." *New York Times*, 4 January 2020. https://www.nytimes.com/2020/01/04/world/australia/fires-scott-morrison.html.
Alexander, Denis. *Creation or Evolution: Do We Have to Choose?* Oxford: Monarch Books, 2014.
Alexander, Denis. *Rebuilding the Matrix*. Oxford: Lion, 2001.
Aristotle. *The Complete Works of Aristotle*, ed. Jonathan Barnes. Princeton, NJ: Princeton University Press, 1984.
Asad, Talal. *Formations of the Secular*. Stanford: Stanford University Press, 2003.
Asad, Talal. *Genealogies of Religion*. Baltimore, MD: John Hopkins University Press, 1993.
Augustine. *Against the Academicians and The Teacher*. trans. Peter King. Indianapolis, IN: Hackett, 1995.
Bacon, Francis. *New Atlantis and The Great Instauration*. Oxford: Wiley-Blackwell, 2012.
Baudrillard, Jean. *Simulacra and Simulation*. Ann Arbor: University of Michigan Press, 2004.
Berger, Peter L. ed. *The Desecularization of the World*. Grand Rapids, MI: Eerdmans, 1999.
Berger, Peter L. and Thomas Luckmann. *The Social Construction of Reality*. New York: Anchor, 1967.
Birol, Fatih. *Key World Energy Statistics 2017*. International Energy Agency. https://web.archive.org/web/20180707173253/https://www.iea.org/publications/freepublications/publication/KeyWorld2017.pdf.
Boersma, Hans. *Seeing God*. Grand Rapids, MI: Eerdmans, 2018.

Bradley, James. "How Australia's Coal Madness Led to Adani." *The Monthly*, April 2019, https://www.themonthly.com.au/issue/2019/april/1554037200/james-bradley/how-australia-s-coal-madness-led-adani.
Brueggemann, Walter. *The Prophetic Imagination*. Minneapolis, MN: Fortress Press, 2001.
Budde, Michael. *The (Magic) Kingdom of God*. Oxford: Westview Press, 1997.
Burnett, David. *The Spirit of Buddhism*. London: Monarch Books, 2003.
Butterfield, Herbert. *Christianity and History*. London: Fontana Books, 1957.
Cahill, Thomas. *How the Irish Saved Civilization*. New York: Anchor Books, 1995.
Cain, Tess. "Scott Morrison's Challenge at Pacific Islands Forum in Tuvalu Is to Deliver on Climate Change." *The Conversation, Australian Broadcasting Corporation News*, 16 August 2019. https://www.abc.net.au/news/2019-08-15/scott-morrisons-pacific-islands-forum-climate-change-challenge/11415832.
Cavanaugh, William T. *Migrations of the Holy*. Grand Rapids, MI: Eerdmans, 2011.
Cavanaugh, William T. *The Myth of Religious Violence*. Oxford: Oxford University Press, 2009.
Chalmers, Alan. *What Is This Thing Called Science?* Berkshire: Open University Press, 2013.
Chryssavgis, John. "Ecumenical Patriarch Bartholomew: Insights into an Orthodox Christian Worldview." *The International Journal of Environmental Studies*, Vol. 64, No. 1 (2007), 9–18.
Cross, Anthony R. and Philip E. Thompson. *Baptist Sacramentalism*. Milton Keynes: Paternoster, 2003.
Cross, Richard. *Duns Scotus*. New York: Oxford University Press, 1999.
Cunningham, Michelle., Luke Van Uffelen, and Mark Chambers. "The Changing Global Market for Australian Coal." *Reserve Bank of Australia Bulletin*, September 2019. https://www.rba.gov.au/publications/bulletin/2019/sep/pdf/the-changing-global-market-for-australian-coal.pdf.
Davies, Paul. *The Mind of God*. London: Penguin, 1993.
Davis, Steven J. and Robert H. Socolow. "Commitment Accounting of CO_2 Emissions." *Environmental Research Letters*, Vol. 9, No. 8 (26 August 2014). https://iopscience.iop.org/article/10.1088/1748-9326/9/8/084018/pdf.
Dawkins, Richard. *The God Delusion*. Boston: Mariner Books, 2008.
De Lubac, Henri. *The Drama of Atheist Humanism*. San Francisco, CA: Ignatius Press, 1995.
Dennett, Daniel C. *Darwin's Dangerous Idea*. Penguin: London, 1996.
Derrida, Jacques. *Of Grammatology*. Baltimore, MD: John Hopkins University, 2016.
Ellul, Jacque. *The Subversion of Christianity*. Grand Rapids, MI: Eerdmas, 1986.
Ellul, Jacque. *The Technological Society*. New York: Vintage, 1964.
Elsworth, Emma. "NSW Bushfires Lead to Deaths of Over a Billion Animals and 'Hundreds of Billions' of Insects, Experts Say." *Australian Broadcasting*

Corporation News, 9 January 2020, https://www.abc.net.au/news/2020-01-09/nsw-bushfires-kill-over-a-billion-animals-experts-say/11854836.
Empiricus, Sextus. *Outlines of Pyrrhonism*. New York: Prometheus Books, 1990.
Epicurus. *The Art of Happiness*. New York: Penguin, 2012.
Farago, Robert. "Nuclear Has Left Its Run Too Late: A Response to Ian Hore-Lacy." *Ethos*, 14 August 2018, http://www.ethos.org.au/online-resources/engage-mail/nuclear-too-late-a-response-to-ian-hore-lacy.
Feyerabend, Paul. *The Tyranny of Science*. Cambridge: Polity Press, 2013.
Finkbeier, Ann K. *The Jasons: The Secret History of Science's Postwar Elite*. New York: Viking, 2006.
Foster, Vivien. and Daron Bedrosyan. *Understanding Carbon Dioxide Emissions from the Global Energy Sector*. The World Bank, briefing paper 85126. http://documents1.worldbank.org/curated/en/873091468155720710/pdf/851260BRI0Live00Box382147B00PUBLIC0.pdf, 2014/5.
Foucault, Michel. *Discipline and Punish: The Birth of the Prison*. London: Penguin, 1991.
Francis. *Laudato Si'*. Città del Vaticano: Libreria Editrice Vaticana, 2015.
Funkenstein, Amos. *Theology and the Scientific Imagination*. Princeton, NJ: Princeton University Press, 1986.
Galbraith, John Kenneth. *The New Industrial State*. Princeton, NJ: Princeton University Press, 2007.
Garnett, Stephen., Brendan Wintle, David Lindenmayer, et al. "Conservation Scientists Are Grieving after the Bushfires – But We Must Not Give Up." *The Conversation*, 21 January 2020. https://theconversation.com/conservation-scientists-are-grieving-after-the-bushfires-but-we-must-not-give-up-130195.
Gaukroger, Stephen. *The Natural and the Human: Science and the Shaping of Modernity 1739–1841*. Oxford: Oxford University Press, 2016.
Gaukroger, Stephen. *Objectivity. A Very Short Introduction*. Oxford: Oxford University Press, 2012.
Gaukroger, Stephen. *The Emergence of a Scientific Culture*. Oxford: Oxford University Press, 2006.
Gaukroger, Stephen. *Francis Bacon*. Cambridge: Cambridge University Press, 2001.
George, Susan. *Hijacking America: How the Religious and Secular Right Changed What Americans Think*. Cambridge: Polity, 2008.
Gerson, Lloyd P. *Ancient Epistemology*. Cambridge: Cambridge University Press, 2008.
Gill, Eric. *Beauty Looks After Herself*. Tacoma, WA: Sophia Perennis et Universalis, 2014.
Goodchild, Philip. *Theology of Money*. Durham, NC: Duke University Press, 2009.
Goodman, Nelson. *Ways of Worldmaking*. Indianapolis, IN: Hackett, 1978.
Gordon, Timothy., Andrew N. Radford, and Stephen D. Simpson. "Grieving Environmental Scientists Need Support." *Science*, Vol. 366, No. 6462 (11 October 2019), 193. doi: 10.1126/science.aaz2422; https://science.sciencemag.org/content/366/6462/193.1.

Grayling, Anthony C. *The God Argument: The Case Against Religion and For Humanism*. London: Bloomsbury, 2013.
Gregory, Brad. *The Unintended Reformation: How a Religious Revolution Secularized Society*. Cambridge, MA: Harvard University Press, 2012.
Griffiths, Tom. "Savage Summer." *Inside Story*, 8 January 2020. https://insidestory.org.au/savage-summer/.
Guardini, Romano. *The End of the Modern World*. Wilmington, DE: ISI Books, 1998.
Guinness, Os. *The Dust of Death: The Sixties Counterculture and How It Changed America Forever*. Wheaton: Crossway Books, 1994.
Hamann, J.G. *Writings on Philosophy and Language*. ed. Kenneth Haynes. Cambridge: Cambridge University Press, 2007.
Hamilton, Clive. *Scorcher: The Dirty Politics of Climate Change*. Melbourne: Black Inc., 2007.
Harmon, Steven R. *Towards Baptist Catholicity*. Milton Keynes: Paternoster, 2006.
Harris, John. "Seeing God: Understanding and Misunderstanding the Dreamtime." *Zadok Perspectives.* Melbourne, Vol. 144 (Spring 2019), 5–9.
Harris, John. *One Blood: 200 Years of Aboriginal Encounter with Christianity: A Story of Hope*. Sydney: Albatross Books, 1990.
Harrison, Peter. *The Territories of Science and Religion*. Chicago, IL: University of Chicago Press, 2015.
Harrison, Peter. *The Fall and Man and the Foundations of Science*. Cambridge: Cambridge University Press, 2007.
Harrison, Peter. "Subduing the Earth: Genesis 1, Early Modern Science, and the Exploitation of Nature." *The Journal of Religion*, Vol. 79, No. 1 (January 1999), 86–109.
Harrison, Peter. *The Bible, Protestantism, and the Rise of Natural Science*. Cambridge: Cambridge University Press, 1998.
Harvey, Barry. *Can These Bones Live? A Catholic Baptist Engagement with Ecclesiology, Hermeneutics, and Social Theory*. Grand Rapids, MI: Brazos Press, 2008.
Hauerwas, Stanley. *After Christendom*. Nashville: Abingdon Press, 1999.
Hayhoe, Katharine. and Andrew Farley. *A Climate for Change*. Nashville, TN: Faith Words, 2009.
Henry, John. *Knowledge Is Power: How Magic, the Government and an Apocalyptic Vision Helped Francis Bacon to Create Modern Science*. London: Icon, 2017.
Henry, John. "Religion and the Scientific Revolution." in Peter Harrison ed., *The Cambridge Companion to Science and Religion*. Cambridge: Cambridge University Press, 2010, 39–58.
Hitchens, Christopher. *God Is Not Great*. London: Atlantic Books, 2008.
Hooykaas, Reijer. *Religion and the Rise of Modern Science*. Vancouver: Regent College Publishing, 2000.

Hore-Lacy, Ian. "Australia's Energy Insanity." *Ethos*, 20 July 2018, http://www.ethos.org.au/online-resources/engage-mail/australias-energy-insanity.

Hore-Lacy, Ian. *Responsible Dominion. A Christian Approach to Sustainable Development*. Vancouver: Regent College Publishing, 2006.

Horn, Allyson. "Election 2019: Why Queensland Turned Its Back on Labor and Helped Scott Morrison to Victory." *Australian Broadcasting Corporation News*, 24 May 2019, https://www.abc.net.au/news/2019-05-19/election-results-how-labor-lost-queensland/11122998.

Houghton, John. *Global Warming. The Complete Briefing*, 5th edition. Cambridge: Cambridge University Press, 2015.

Hunter, James Davison. and Os Guinness, eds. *Articles of Faith, Articles of Peace: The Religious Liberty Clauses and the American Public Philosophy*. Washington, DC: The Brookings Institute, 1990.

Hutchings, Maggie. *I Saw Pete and Pete Saw Me*. Mulgrave, Victoria, Australia: Affirm Press, 2020.

Illich, Ivan. "Guarding the Eye in the Age of Show." *RES: Anthropology and Aesthetics*, No. 28 (Autumn 1995), 47–61. http://www.davidtinapple.com/illich/2001_guarding_the_eye.PDF.

Inhofe, James. *The Greatest Hoax: How the Global Warming Conspiracy Threatens Your Future*. USA: WorldNetDaily Books, 2012.

Inhofe, James. US Senate Floor Speech, "Catastrophic Global Warming Alarmism Not Based on Objective Science," delivered on 28 July 2003. Quotes from this speech are cited here: https://www.nytimes.com/2003/08/05/science/politics-reasserts-itself-in-the-debate-over-climate-change-and-its-hazards.html?pagewanted=all&src=pm.

Jackson, Tim. *Prosperity Without Growth*. London: Routledge, 2017.

Jenkins, Philip. *The New Face of Christianity: Believing the Bible in the Global South*. Oxford: Oxford University Press, 2006.

John the Scot. *Periphyseon*. trans. Myra L. Uhlfelder, Eugene, OR: Wipf & Stock, 2011.

Johnson, Monte. and Catherine Wilson. "Lucretius and the History of Science." in Stuart Gillespie and Philip Hardie eds., *The Cambridge Companion to Lucretius*. Cambridge: Cambridge University Press, 2007, 131–148.

Kahneman, Daniel. On the radio program, "Hidden Brain" hosted by Shankar Veantam, "Daniel Kahneman on Misery, Memory, and Our Understanding of the Mind." *National Public Radio, USA*, 12 March 2018, https://www.npr.org/2018/03/12/592986190/daniel-kahneman-on-misery-memory-and-our-understanding-of-the-mind.

Kahneman, Daniel., Paul Slovic, and Amos Tversky, eds. *Judgment under Uncertainty: Heuristics and Biases*. Cambridge: Cambridge University Press, 2019.

Keele, Rondo. *Ockham Explained*. Chicago, IL: Open Court, 2010.

Khadem, Nassim. "Ross Garnaut's Climate Change Prediction Is Coming True and It's Going to Cost Australia Billions, Experts Warn." *Australian*

Broadcast Corporation News, 8 January 2020, https://www.abc.net.au/news/2020-01-08/economic-bushfires-billions-ross-garnaut-climate-change/11848388.

Kierkegaard, Søren. *Concluding Unscientific Postscript to the Philosophical Fragments*. Princeton, NJ: Princeton University Press, 1992.

Kirk, G.S., J. E. Raven, and M. Schofield. *The Presocratic Philosophers*, 2nd edition. Cambridge: Cambridge University Press, 2007.

Klein, Naomi. *On Fire*. London: Penguin, 2019.

Krauss, Lawrence M. *The Greatest Story Ever Told... So Far. Why Are We Here?* New York: Simon & Schuster, 2017.

Latour, Bruno. *Science in Action*. Cambridge, MA: Harvard University Press, 2011.

LeVasseur, Todd. and Anna Peterson. eds. *Religion and Ecological Crisis: The "Lynn White Thesis" at Fifty*. London: Routledge, 2018.

Liethart, Peter. *Against Christianity*. Moscow, ID: Canon Press, 2003.

Lightman, Bernard. ed. *Rethinking History, Science, and Religion: An Exploration of Conflict and the Complexity Principle*. Pittsburgh, PA: University of Pittsburgh, 2019.

Lightman, Bernard. ed. *Victorian Popularizers of Science: Designing Nature for New Audiences*. Chicago, IL: University of Chicago Press, 2011.

Lightman, Bernard. and Michael S. Reidy. eds. *The Age of Scientific Naturalism: Tyndall and His Contemporaries*. Pittsburgh, PA: University of Pittsburgh Press, 2014.

Lindsey, Rebecca. "Climate Change: Atmospheric Carbon Dioxide." National Oceanic and Atmospheric Administration, U.S. Department of Commerce. Climate.gov, 19 September 2019, https://www.climate.gov/news-features/understanding-climate/climate-change-atmospheric-carbon-dioxide.

Lovelock, James. *Gaia: A New Look at Life on Earth*. Oxford: Oxford University Press, 2016.

Low, Mary. *Cherish the Earth: Reflections on a Living Planet*. Glasgow: Wild Goose Publications, 2003.

Low, Mary. *Celtic Christianity and Nature*. Edinburgh: Edinburgh University Press, 1997.

Lucretius. *On the Nature of the Universe*. Oxford: Oxford University Press, 1999.

MacIntyre, Alasdair. *After Virtue*, 3rd edition. Notre Dame, IN: University of Notre Dame Press, 2007.

Maslin, Mark. *Climate Change: A Very Short Introduction*, 3rd edition. Oxford: Oxford University Press, 2014.

McGrath, Alister E. *The Re-Enchantment of Nature*. London: Hodder & Stoughton, 2003.

McNeill, J.R. and Peter Engelke. *The Great Acceleration*. Cambridge, MA: Harvard University Press, 2014.

Meek, Donald E. *The Quest for Celtic Christianity*. Fife: Handsel, 2000.

Menn, Stephen. "Aristotle and Plato on God as Nous and as the Good." *The Review of Metaphysics*, Vol. 45, No. 3 (March 1992), 543–573.
Merchant, Carolyn. *The Death of Nature*. New York: HarperOne, 1990.
Merritt, Jonathan. *Green Like God*. New York: Faithwords, 2010.
Merzian, Richie. "Taking Way Too Much Credit." *The Australia Institute Briefing Note*, May 2019. https://www.tai.org.au/sites/default/files/P645%20 Taking%20way%20too%20much%20credit%20%5BWEB%5D.pdf.
Milbank, John. *Theology and Social Theory*, 2nd edition. Oxford: Blackwell, 2008.
Moyers, Bill. *Welcome to Doomsday*. New York: New York Review of Books, 2006.
Nagel, Ernst. and James R. Newman. *Gödel's Proof*. New York: New York University Press, 2001.
Nietzsche, Friedrich. *The Gay Science*. Cambridge: Cambridge University Press, 2019.
Noll, Mark A. *America's God*. Oxford: Oxford University Press, 2002.
Nongbri, Brent. *Before Religion: A History of a Modern Concept*. New Haven, CT: Yale University Press, 2013.
Northcott, Michael. "Lynn White Jr. Right and Wrong: The Anti-Ecological Character of Latin Christianity and the Pro-Ecological Turn of Protestantism." in Todd LeVasseur and Anna Peterson eds., *Religion and Ecological Crisis*. London: Routledge, 2017, 61–74.
Northcott, Michael. *A Political Theology of Climate Change*. London: SPCK, 2014.
Numbers, Ronald L. ed. *Galileo Goes to Jail and Other Myths about Science and Religion*. Cambridge, MA: Harvard University Press, 2009.
Oreskes, Naomi. and Erik M. Conway. *Merchants of Doubt*. New York: Bloomsbury, 2010.
Pangarta, Ken Lechleitner. "Central Australian Aboriginal People's Worldview on God/Altijirra." *Zadok Perspectives*, Melbourne, Vol. 144 (Spring 2019), 3–4.
Pasnau, Robert. *After Certainty: A History of Our Epistemic Ideals and Illusions*. Oxford: Oxford University Press, 2017.
Pasnau, Robert. *Metaphysical Themes 1274–1671*. Oxford: Oxford University Press, 2011.
Polanyi, Michael. *Personal Knowledge*. Chicago, IL: University of Chicago Press, 1974.
Posner, Sarah. *Unholy: Why White Evangelicals Worship at the Alter of Donald Trump*. New York: Random House, 2020.
Price, Simon. *Religion of the Ancient Greeks*. Cambridge: Cambridge University Press, 2011.
Quiggin, John. "Explaining Adani." *The Conversation*, 3 June 2019. https://theconversation.com/explaining-adani-why-would-a-billionaire-persist-with-a-mine-that-will-probably-lose-money-117682.
Reynolds, Henry. *The Other Side of Frontier*. Ringwood: Penguin, 1982.

Roochnik, David. *Retrieving Aristotle in an Age of Crisis*. New York: State University of New York Press, 2013.
Ross, Chanon. *Gifts Glittering and Poisoned: Spectacle, Empire and Metaphysics*. Eugene, OR: Cascade, 2014.
Rowland, Tracey. *Ratzinger's Faith*. Oxford: Oxford University Press, 2008.
Rowley, Charles D. *The Destruction of Aboriginal Society*. Ringwood: Penguin, 1972.
Rubenstein, Richard. *Aristotle's Children*. New York: Harcourt, 2003.
Schaeffer, Francis. *Pollution and the Death of Man*. Wheaton, IL: Crossway Books, 1992.
Schaeffer, Francis. *How Shall We Then Live?* Wheaton, IL: Crossway Books, 1976.
Schmemann, Alexander. *For the Life of the World*. New York: St Vladimir's Seminary Press, 1973.
Schmitt, Charles B. *Aristotle and the Renaissance*. Cambridge, MA: Harvard University Press, 1983.
Serreze, Mark C. *Brave New Arctic: The Untold Story of the Melting North*. Princeton, NJ: Princeton University Press, 2018.
Shapin, Steven. *The Scientific Revolution*, 2nd edition. Chicago, IL: University of Chicago Press, 2018.
Shapin, Steven. *Never Pure: Historical Studies of Science as if It Was Produced by People with Bodies, Situated in Time, Space, Culture, and Society, and Struggling for Credibility and Authority*. Baltimore, MD: John Hopkins University Press, 2010.
Shapin, Steven. and Simon Schaffer. *Leviathan and the Air-Pump: Hobbes, Boyle, and the Experimental Life*. Princeton, NJ: Princeton University Press, 2011.
Sherwood, S., Webb, M.J., Annan, J.D., et al. "An Assessment of Earth's Climate Sensitivity Using Multiple Lines of Evidence." World Climate Research Program, Geneva, Switzerland, 15 July 2020, https://climateextremes.org.au/wpcontent/uploads/2020/07/WCRP_ECS_Final_manuscript_2019RG000678R_FINAL_200720.pdf.
Smith, Christian. *American Evangelicalism: Embattled and Thriving*. Chicago, IL: University of Chicago Press, 1998.
Smith, Wilfred Cantwell. *The Meaning and End of Religion*. Minneapolis, MN: Fortress, 1991.
Spencer, Nick. *Atheists: The Origin of the Species*. London: Bloomsbury, 2014.
Spencer, Nick. *Darwin and God*. London: SPCK, 2009.
Stott, John. *Issues Facing Christians Today*. UK: Marshall, 1984.
Subramanian, Meera. *Finding Middle Ground: Conversations across America about Climate Change*. USA: Independently Published, 2019.
Subramanian, Meera. "Generation Climate: Can Young Evangelicals Change the Climate Debate?" *Inside Climate News*, 21 November 2018. https://insideclimatenews.org/news/21112018/evangelicals-climate-change-action-creation-care-wheaton-college-millennials-yeca.

Taylor, Charles. *A Secular Age*. Cambridge, MA: Harvard University Press, 2007.
Thearle, M. John. "The Rise and Fall of Phrenology in Australia." *Australian and New Zealand Journal of Psychiatry*, Vol. 27, No. 3 (1993), 518–525.
Thomas, Keith. *Religion and the Decline of Magic*. London: Penguin, 1991.
Thunberg, Greta., Svante Thunberg, Malena Ernman, and Beata Ernman. *Our House Is on Fire*. London: Penguin, 2020.
Tillich, Paul. *A History of Christian Thought*. New York: Simon and Schuster, 1968.
Tyson, Paul. *Seven Brief Lessons on Magic*. Eugene: Cascade, 2019.
Veldman, Robin Globus. *The Gospel of Climate Change Skepticism: Why Evangelical Christians Oppose Action on Climate Change*. Berkeley: University of California Press, 2019.
Vidal, John. and Damian Carrington. "Is Climate Change to Blame for Typhoon Haiyan?" *The Guardian*, 13 November 2013, https://www.theguardian.com/world/2013/nov/12/typhoon-haiyan-climate-change-blame-philippines.
Vince, Gaia. and Gabriel Gatehouse. "Ecological Greif." *BBC World Service*, 20 January 2020, https://www.bbc.co.uk/sounds/play/w3csy6d0.
Virilio, Paul. *The Great Accelerator*. Cambridge: Polity, 2012.
Webber, Robert E. *Ancient-Future Faith*. Grand Rapids, MI: Baker Academic, 1999.
Weber, Max. *The Protestant Ethic and the Spirit of Capitalism*. London: Unwin, 1985.
Webster, Charles. *Paracelsus: Medicine, Magic and Mission at the End of Time*. New Haven, CT: Yale University Press, 2008.
Webster, Charles. *The Great Instauration*. London: Duckworth, 1975.
Werner, Joel. "Wildfires with Wild Numbers: Fact Checking a Catastrophe." *Science Friction*, Australian Broadcasting Corporation, 16 February 2020, https://www.abc.net.au/radionational/programs/sciencefriction/16.1-fact-checking-the-fire-season/11962758.
White Jr., Lynn. "The Historical Roots of Our Ecological Crisis." *Science*, Vol. 155, No. 3767 (10 March 1967), 1203–1207. https://www.cmu.ca/faculty/gmatties/lynnwhiterootsofcrisis.pdf.
White Jr., Lynn. *Medieval Technology and Social Change*, Oxford: Oxford University Press, 1962.
White Jr., Lynn. "Christian Myth and Christian History." *Journal of the History of Ideas*. Vol. 3, No. 2 (April 1942), 145–158.
White, Robert M. "Oceans and Climate – Introduction." *Oceanus*, Vol. 21 (1978), 2–3.
Yannaras, Christos. *Against Religion*. Brookline: Holy Cross Orthodox Press, 2013.
Yannaras, Christos. *The Freedom of Morality*. New York: St Vladimir's Seminary Press, 1984.

Yunkaporta, Tyson. *Sand Talk: How Indigenous Thinking Can Save the World.* Melbourne: Text, 2019.

Zizioulas, John. "Preserving God's Creation: Three Lectures on Theology and Ecology." *King's Theological Review,* Vol. 12 (1989), 1–5; *King's Theological Review,* Vol. 12 (1989), 41–45; *King's Theological Review,* Vol. 13 (1990), 1–5.

Index

Abelard, Peter 39, 40, 43
Adam 59–62, 70
Alexander, Denis 108
American Way 96, 99, 104, 107, 108, 117
Anselm 65
anti-Darwinism 96
apocalypse 115–6
Aquinas 44, 45, 64
Aristotle 25–9, 40–8, 53, 58, 63, 64, 122
atomism 7, 30, 31, 38, 46–8, 71
Augustine 3, 39, 43–5, 59, 62
Averroes (Ibn Rushd) 44

Bacon, Francis 24, 34, 47, 54, 66, 68, 69
Bartholomew, Ecumenical Patriarch 85
Berger, Peter L. 125
Boersma, Hans 87
Boethius 43
Boyle, Robert 101
Brueggemann, Walter 87
Bush, George H. 104
Bush, George W. 102, 118
bushfires (Australia 2019–2020) 11–3
Butterfield, Herbert 98

Cahill, Thomas 84
Calvinism 45–6
Carter, Jimmy 103

climate change denial 56, 99, 100, 103, 104, 116
climate change scepticism 4, 108, 116, 117, 124
climate science scepticism 6, 106, 108
climate science 5–6, 86, 102–7, 110
Conflict Thesis (science and religion) 16
Coulter, Ann 105
creation care (Evangelical) 15, 35, 97, 99–102, 120
Creation Science 96
culture wars 16, 101, 118, 123

Darwin, Charles 34
De Lubac, Henri 81–3
Deism 48, 67, 101
Democritus 30, 46–9
Descartes, René 41, 56, 64, 65
Dobson, James 99
dominion 8, 52, 60, 68–71, 80, 105, 108–12, 123

eco-theology, Celtic 84–5
eco-theology, Christian Indigenous 87–90
eco-theology, Evangelical 95–112
eco-theology, Main-line 90, 91
eco-theology, Orthodox 85–7
eco-theology, Roman Catholic 79–83
embattled Evangelicals 116–7
embattled Progressive 117–9

empiricism 46, 55, 58, 65, 66
Epicurus 30
Eriugena, John Scotus 84–5
eschatology 68, 71, 98
essence 32, 33, 39–42, 46, 48, 53–5, 61, 62
Evangelical Climate Initiative 102
Evangelical Environment Network 102
Evangelical greening 99, 100, 102, 104, 106, 108
existence 30, 31, 33, 39, 41, 46, 61–3

fall (the) 24, 59–69, 80
Falwell, Jerry 97, 99
Faraday, Michael 101
Feuerbach, Ludwig 34
first philosophy (Theology A) 7, 25–9, 38–41, 46, 51, 71, 102, 122
form 7, 39–42, 46, 61, 62, 85
Francis (Pope) 79–83
Francis (of Assisi) 79
Funkenstein, Amos 56

Galilei, Galileo 41, 53–6, 64, 65, 69
Gassendi, Pierre 47
George, Susan 118–9
Gore, Al 99, 103, 105
Grotius, Hugo 121
Guardini, Romano 80–3

Hagee, John 106
Harris, John 89
Harrison, Peter 2–4, 58
Hayhoe, Katherine 105–8
Hegel, G.W.F. 81, 90, 101
hermeneutic transition 100–1
homelessness 124
Hooykaas, Reijer 58
Houghton, John 99, 109
Hume, David 48
hylomorphism 46–8

Illich, Ivan 54, 63
Inhofe, James 106, 108
intellectualist (medieval) 45
Intergovernmental Panel on Climate Change (IPCC) 103, 105

Jenkins, Philip 87
Jesus 34, 68, 106

Johnson, Lyndon B. 103

Kahneman, Daniel 4, 124
Kant, Immanuel 90, 101
Kyoto Protocol 105

Lewis, C.S. 108
life-world 4, 7, 14–5, 24, 28, 46, 120–6
light (natural and divine) 64–7
Lightman, Bernard 16
Lovelock, James 32
Low, Mary 85
Lucretius 30–1
Luther, Martin 58, 64, 65

MacIntyre, Alasdair 55, 66
matter (modern atomist, and hylomorphic medieval) 31, 40–1, 46–9, 71
McGrath, Alister 108
Merritt, Jonathan 99, 100, 102, 121
metaphysics 55, 66
Milbank, John 2–4
modern philosophy 26
Moral Majority 97, 99, 117

National Oceanic and Atmospheric Administration 103
natural magic 85
naturalism 30, 31, 96, 98, 101, 111, 122
Nature and Culture 27, 42–3
necessitism 44–5
Newton, Isaac 48, 53, 54
Nierenberg, Bill 104
nominalism 7, 38–43, 45, 71, 86, 107
Northcott, Michael 122–3
nuclear energy 109

Obama, Barack 104
Ockham, William of 40, 44, 46, 64
out-groups 119–21

Pangarta, Ken Lechleitner 89
Pasnau, Robert 39, 53
Plato 40, 41, 47
political theology 83, 115–23
politics 4–6, 11–3, 115–9, 121–6
power (knowledge as) 52–6
pragmatism 66

Index

Protestants 41, 48, 57, 58, 63–8, 101
Pufendorf, Samuel von 121
Pyrrho 30

rationalism 65
Reagan, Ronald 103
realism (medieval) 39
realism (modern) 42–3
Reformation 41, 45, 57, 58, 64, 65
religion 82
Religious Right 4–6, 99–100, 104, 105, 108, 115–9
Resourcement 81
responsible dominion 108–10
Royal Society of London 47, 58
Rubenstein, Richard 41

sceptics 64, 66
Schaeffer, Francis 97–102
Schelling, Thomas 104
Schleiermacher, Friedrich 90
Schmemann, Alexander 86
Schmitt, Charles 64
science and religion 1–4
Scotus, Duns 40, 44, 45
secular theology 56–7
secularization 52, 125
Shakespeare 68
social Darwinism 88
sociology of knowledge 4, 14, 15
stewardship (Evangelical) 98, 106, 107
Stott, John 109
Strauss, David 34, 90

Taylor, Charles 72
theologies of nature: Epicurean 29–32; Animist 32; Christian 32–3; Eastern 33
theology (as first philosophy) 25–9
theology (causal of climate change) 13–7
Theology A (first philosophy) and Theology B (religious/systematic theology) 51–2
theophanic 85
Thunberg, Greta 119
time 23–5
Tolkien, J.R.R. 108

Veldman, Robin Globus 97, 100, 115, 119
voluntarism 43–6, 49, 69, 71, 80, 84, 86, 107

Webber, Robert 87
Webster, Charles 58
Wesley, John 90
White Evangelicals 110–2
White House 102–5
White Jr., Lynn 13–5, 23–5, 125
White, Andrew Dickson 34

Young Earth Creationism 118, 120
Yunkaporta, Tyson 88

Zizioulas, John 86

For Product Safety Concerns and Information please contact our EU representative GPSR@taylorandfrancis.com
Taylor & Francis Verlag GmbH, Kaufingerstraße 24, 80331 München, Germany

www.ingramcontent.com/pod-product-compliance
Lightning Source LLC
Chambersburg PA
CBHW051750230426
43670CB00012B/2229